SpringerBriefs in Earth Sciences

More information about this series at http://www.springer.com/series/8897

Erdal Yiğit

Atmospheric and Space Sciences: Neutral Atmospheres

Volume 1

 Springer

Erdal Yiğit
Department of Physics and Astronomy
George Mason University
Fairfax, VA
USA

ISSN 2191-5369   ISSN 2191-5377   (electronic)
SpringerBriefs in Earth Sciences
ISBN 978-3-319-21580-8   ISBN 978-3-319-21581-5   (eBook)
DOI 10.1007/978-3-319-21581-5

Library of Congress Control Number: 2015945124

Springer Cham Heidelberg New York Dordrecht London

Printed on acid-free paper

Springer International Publishing AG Switzerland is part of Springer Science+Business Media
(www.springer.com)

*Dedicated to my beloved parents
İhsan & Nazlı Yiğit and to all children
of the universe.*

Foreword

We are witnessing a quiet revolution in atmospheric science. First, an increasing number of scientific papers on the subject include phrases like "coupling", "interaction", "influence" in their titles. This reflects the transition from exploring isolated atmospheric phenomena, processes, and layers to understanding their relations, and our environment as a whole system. Second, we have learned a lot about atmospheres of planets and satellites of the Solar System in addition to our own native Earth. Discoveries over the past decade of hundreds, and now thousands, of extrasolar planets bring to our attention even more exotic atmospheric environments. How can we understand them? Do we have to study one planet after another? Fortunately, not. We believe that atmospheric processes on other planets are governed by the same physical laws as on Earth, unless, of course, we find their violation. This book gives a concise snapshot of these laws, concepts, mathematical tools, and approaches developed by atmospheric and space scientists for Earth. As more young researchers enter the field, they are challenged with how to shorten their way to the cutting-edge problems. The author, who himself is a young and productive scientist with whom I have had the privilege to work with, knows how to bridge this gap: directly from basics to the forefront research articles. Established atmospheric, planetary, and space scientists extending their research interests will find this book useful and interesting too.

Göttingen   Alexander S. Medvedev
May 2015

Preface

My goal with the publication of this two-volume "Atmospheric and Space Sciences" Springer Briefs series is to contribute toward bridging the gap between the scientific disciplines of meteorology, aeronomy (or space science), and planetary science and, in particular, provide a whole atmosphere systems science approach to Earth's atmosphere and plasma environment. As such, this environment is a multi-component, complex system that is governed by a nonlinear interplay of physical and chemical processes. Technically, one could speak of the plasma physics, chemistry, electrodynamics, modeling and observation, and hydrodynamics of the atmosphere–ionosphere system. There are separate seminal textbooks out there on each of these aspects of the atmosphere in great detail. My text is not meant to present a comprehensive overview of the entire field. Nevertheless, these volumes are thought to provide a basic and practical introduction to the fundamental physics of atmospheric and ionospheric processes. Specifically, the two volumes of the "Atmospheric and Space Sciences" are

- Volume 1—Neutral Atmospheres
- Volume 2—Plasma Environment

The first volume focuses on the fundamentals of terrestrial and planetary neutral atmospheres. The second volume focuses more on the plasma basics of the terrestrial and planetary atmospheres and, in particular, the thermosphere–ionosphere. In both volumes, some selected research topics are included and relevance to planetary science is highlighted where appropriate.

Overall, these volumes serve as a concise introduction to the basics of atmospheric and space sciences and they highlight some current research activities in these fields. In particular, the physics of internal gravity waves and vertical coupling have extensively been discussed throughout the text because of the interdisciplinary nature of these subjects. Many details have been left out for the interested readers to check in detail themselves. Scientific problems whose solution significantly involves an understanding of Earth's atmosphere as a whole have brought scientists from different subfields together. Here, I would like to emphasize that often the expressions of "in the atmosphere" and "in planetary atmospheres" have been used interchangeably in the text.

These volumes are accessible to a broad range of audience. Especially, undergraduate and graduate students, for example in physics, mathematics, and Earth sciences, can greatly benefit from the text. For senior undergraduate science students, these volumes could be very helpful to learn the basics of atmospheric physics. Early career geoscience researchers can use the books to review various topics of interest and develop new ideas. Overall, I envisage that anyone with some appreciation for basic mathematics and geosciences will enjoy reading these volumes.

Fairfax, VA   Erdal Yiğit
May 2015

Acknowledgments

First of all, I would like to thank all the authors of the publications that are cited in the references of this book for having inspired me and indirectly helped me write this interdisciplinary book.

I would like to express my appreciation to Petra van Steenbergen, Springer Senior Publishing Editor for Earth Sciences and Geography, for supporting my book idea and organizing the peer-review process and all the related publication processes. Hermine Vlomans, Assistant to Mrs. van Steenbergen, has greatly helped me clarify issues related to manuscript preparation and formatting.

My conversations on dialectic materialism with Dr. Alexander Kutepov at NASA Goddard Space Flight Center have been extremely valuable and have made me revisit some of the ideas presented in this book and approach some of the topics from a critical point of view.

I have had many fruitful conversations with Prof. Michael Summers at the George Mason University on planetary science. These dialogues were very helpful in shaping my perspective on writing in sciences for a general public.

With his scientifically honest and rigorous way, Dr. Alexander S. Medvedev has been a source of motivation and enthusiasm for me since the early stages of my Ph.D.

I am grateful for my parents' spiritual support at all stages of my scientific endeavor and especially for the excitement and support they have given during the design my book.

Contents

Acronyms

BC	Before Christ
CHAMP	CHAllenging Minisatellite Payload
CME	Coronal Mass Ejection
DMSP	Defense Meteorological Satellite Program
EDVAC	Electronic Discrete Variable Automatic Computer
ENIAC	Electronic Numerical Integrator and Calculator
ESA	European Space Agency
EUV	Extra Ultraviolet
GAIA	Ground-to-topside Model of Atmosphere and Ionosphere for Aeronomy
GCM	General Circulation Model
GRACE	Gravity Recovery and Climate Experiment
GW	Gravity Wave
HIRDLS	High Resolution Dynamics Limb Sounder
IAGA	International Association of Geophysics and Aeronomy
IAU	International Astronomical Union
ICMA	International Commission in the Middle Atmosphere
IUGG	International Union of Geodesy and Geophysics
MAVEN	Mars Atmosphere and Volatile EvolutioN
MHD	Magnetohydrodynamic(s)
NASA	National Aeronautics and Space Administration
ROSMIC	Role Of the Sun and the Middle atmosphere/thermosphere/ionosphere In Climate
SCOSTEP	Scientific Community on Solar-Terrestrial Physics
SOHO	Solar and Heliospheric Observatory
SSW	Sudden Stratospheric Warming
TEM	Transformed Eulerian Mean
TIMED	Thermosphere Ionosphere Mesosphere Energetics and Dynamics
UT	Universal Time
UV	Ultra-Violet
VarSITI	Variability of the Sun and Its Terrestrial Impact

Chapter 1
A Brief Overview of Atmospheric and Space Sciences

Toward a Unified Approach

It must be acknowledged that all the sciences are so closely interconnected that it is much easier to learn them all together than to separate one from the other. If, therefore, someone seriously wishes to investigate the truth of things, he ought not to select one science in particular, for they are all interconnected and interdependent.

—Descartes (17th century A.D.)

Abstract This chapter is aimed to provide a brief overview of atmospheric, space and planetary sciences. The significance of Earth as a habitable planet is mentioned, discussing basic concepts of habitability. A unification in these sciences is discussed based on the common scientific approaches and methodologies. A brief summary of various international organizations of atmosphere-ionosphere science and the associated activities are presented. Finally, some recent planetary missions, such as MAVEN and New Horizon, related to Mars and Pluto are highlighted.

Keywords Earth · Planets · Habitability · Weather · Space weather · Atmosphere-ionosphere · Space science · Planetary science and missions

1.1 Earth as a Special Habitat

In a broader context, atmospheric and planetary sciences focus on the formation, past evolution and the current and future state of Earth and other planets within and outside the Solar System. The exoplanets (or extra-solar planets), that is, planets outside our Solar System are being increasingly detected. So far, more than 1800 exoplanets have been detected[1] and the number is continuously increasing. For example, this number was about 300 in 2009. To our astonishment, there are indications that some of them are Earth-like. In the Solar System, the inner terrestrial planets, Mercury, Venus, Earth, and Mars have a solid land surface, while the outer planets, such as,

---

[1] http://www.exoplanet.eu/.

Fig. 1.1 The Sun and the Solar System planets: Terrestrial inner planets Mercury, Venus, Earth, and Mars, and the outer planets, the gas giants Jupiter and Saturn, Uranus, and Neptune. The sizes of the planets in the figure reflect approximately their relative size with respect to each other. The distances between the planets are not scaled. Pluto is on the far right side. Credit: Lunar and Planetary Institute

Jupiter and Saturn called "gas giants", have no solid surface although they are much larger than the terrestrial inner planets as illustrated in Fig. 1.1, where their sizes in the figure have been scaled with respect to the actual planetary radii. Jupiter is the largest planet in the Solar System and Saturn is immediately recognizable owing to its spectacular rings. In 2006, the International Astronomical Union (IAU) determined that Pluto cannot be considered a planet anymore because it does not sufficiently "clear the neighbourhood around its orbit" and ultimately classified it as a "dwarf planet". Overall our Solar System accommodates a great deal of geophysical and astrophysical diversity.

So, how long has our planet been around for? Figure 1.2 puts Earth's age, which is about 4.5 billion years, in the context of astronomical time scales in years. To the current state of knowledge, our universe was formed about 13.8 billion years ago by the "Big Bang" and has been expanding since then. Because the speed of light is limited ($c \approx 3 \times 10^8$ m s^{-1}), we can only detect the light that has reached us since the Big Bang. Thus, we can only study the "observable universe". Our Sun was born about 5 billion years ago and the planet Earth (and most probably all the other planets in the Solar System) were formed about 500 million years after that. According to archeological excavations, scientists predict that dinosaurs appeared around 230 millions years ago. Homo sapiens sapiens, the species of the modern human, is thought to have evolved about 100,000 years ago. Historical records suggest that the Egyptians have established the first empire about 3000 years BC. Also, around 4–3 millenia BC, Sumerians have invented writing. The industrial revolution, which enabled various modern manufacturing processes and thus led to technological advancements, took place in the 18–19 century. Since then, technological and scientific knowledge are rapidly advancing.

The concepts of *habitability* and *life* are closely interconnected. While the discovery of new planets outside our Solar System is an exciting endeavor, the question of whether there is or could have been any life on these planets is another level of great excitement. So, are there any habitable places out there? Habitable means suitable to

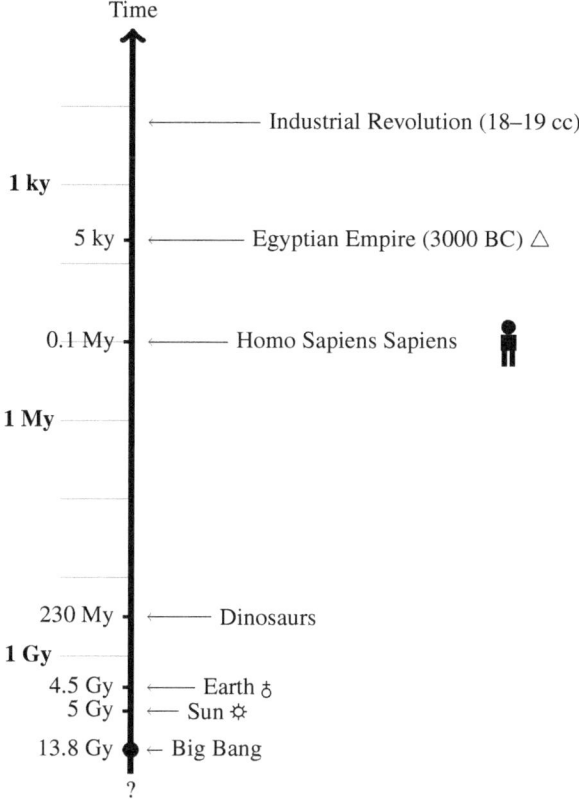

Fig. 1.2   The age of Earth in the context of astronomical time scales, where 1 Gy= 10^9, 1 My = 10^6, and 1 ky = 10^3 years. The specified times on the *left hand side* of the time axis indicate approximately how many years ago a specific event took place with respect to today. For example, the universe is thought to have formed about 13.5 billion years ago by the "Big Bang". The question mark suggests that the current state of science has no feasible explanation for what was before the Big Bang. Sun and Earth have formed approximately 5 and 4.5 billion years ago, respectively. The time axis is approximately logarithmic, where each *gray* long tick mark denotes an order of logarithmic magnitude

live in. There are a number of astronomical factors that influence habitability, such as, the distance from a host star. However, the question of habitability is more complex than simply for the planet to be at the right distance from its host star for liquid water to form on the surface. The magnetic field environment around the planet, various geodynamical processes, and radiative balance (reflection and absorption of stellar energy) are significant factors (Lammer et al. 2009). In the presence of a planetary magnetic field, the atmosphere of a planet is better protected from the solar wind, which is the flow of energetic particles from the host star (Dehant et al. 2007). There-fore, a host star does not only provide the energy for life but it can also take it, if an existing life form is not adequately protected from its hard radiation. Astronomers have introduced the concept of a habitable zone. It is a spherical hypothetical region centered around a star where a given planet with an atmosphere has the right conditions to hold liquid water at a given time (Huang 1959; Lammer et al. 2009).

Among all the known planets, Earth is truly a very special planet. It has provided optimal biological and physical circumstances for the formation of life. To our current state of knowledge, Earth is the only planet that offers natural habitability. Any other planetary system in the Solar System would have to artificially be transformed into a habitable environment for life to exist. There are substantial technological challenges in this transformation. There is ongoing research on the formation and sustainability of small-scale habitable ecosystems.

What makes Earth a habitable planet? The right distance from the Sun, a well-developed plasma environment, optimal radiative conditions, and a breathable atmosphere are some of the conditions that we can think of, in addition to water, the liquid form of H_2O. No other planet in the Solar System has the appropriate atmospheric conditions for the type of life we are familiar with to form. The human body has "evolved" apparently to survive only in Earth's atmosphere. Therefore, understanding the current state and evolution of the atmosphere is of great interest from a basic perspective of human life. Of course, it is scientifically possible that there could be other forms of life that are yet to be discovered. In other words, to exclude the possibility of the existence of life in the universe other than on Earth is scientifically not fully justified.

1.2 Atmospheric and Space Sciences: Toward Unification

One of the major topics of atmospheric sciences is Earth's weather, which is associated with short-term variations in the atmosphere. Weather systems in the lower atmosphere have great impact on the daily life, agriculture, and technology. Often we plan our personal and professional life according to the weather forecast. The concept of "weather" concerns the short-period changes in various meteorological processes. Severe weather conditions can be dangerous and weather watches are generated around the world for the announcement of a potential severe weather. Weather warnings announce that severe weather conditions are imminent. Predictions of meteorological processes is of great challenge because of the complexity and nonlinear nature of weather. Often turbulence in nature is thought to be responsible for the unpredictable character of weather. Rapid localized variations can occur in weather systems. Such complexity is not limited to the lower atmosphere. The geospace environment is largely influenced by the magnetosphere and Sun. Geomagnetic and solar effects on Earth's atmosphere and the resulting short-term variations are broadly called "space weather". This expression has been frequently used in recent times in order to emphasize the variable nature of the upper atmosphere in the context of Sun-Earth coupling (e.g., Blanch et al 2013).

Space weather and lower atmospheric weather deal with similar aspects of natural complexity and fundamental geophysical problems. Their descriptions are both based on similar fundamental physical laws. Consider some general questions that could be of common interest to the scientists in these disciplines:

- What is the impact of the solar activity on the system?
- What are the fundamental processes that are responsible for coupling within the system?
- How can changes be predicted?
- What are the consequences/effects of variability within the system?
- What is a good strategy to embark on the fundamental questions?
- How can forecasting capabilities be improved for weather?

The expression "system" above refers generally to any component of the atmosphere and ionosphere in terrestrial and planetary atmospheres.

Meteorology and space sciences are historically two distinct disciplines. They have their own organizations, research journals, set of techniques, and scientific "jargon". Often, there has been insufficient communication between these two communities, even in the subjects that could have immensely benefited from interdisciplinary efforts and contributions. One of the main reasons for this division has been the prevailing overall perception that their scientific questions and foci are fundamentally different.

However, the scientific community gradually realizes that these disciplines have many common problems and challenges and they could potentially benefit from each others' scientific methods and applications. In particular, the desire to understand various coupling processes in Earth's whole atmosphere-ionosphere system has made it clear that fundamental science problems can better be investigated if a systems science approach is implemented. That is, an interaction between the lower atmosphere scientists and space scientists or aeronomists is crucial for atmospheric and space research to progress.

1.3   International Activity in Atmosphere-Ionosphere Science

Nowadays, there is an increasing amount of interaction between meteorologists and space scientists around the world. But at what platforms can researchers around the world actually interact? How are atmosphere and ionosphere sciences represented? International geoscience organizations have undisputedly played a pivotal role in fostering a better communication between scientists. Around the world there is a large number of international organizations, conferences, and symposia that are dedicated to encourage a whole atmosphere-ionosphere research and the scientific communities benefit from such interactions greatly. For example, Division II—*Aeronomic Phenomena* of the International Association of Geomagnetism and Aeronomy[2] (IAGA)

---

[2]http://www.iugg.org/IAGA/iaga_pages/science/division_2.htm.

Fig. 1.3 International
Associations that form the
International Union of
Geodesy and Geophysics
(IUGG)

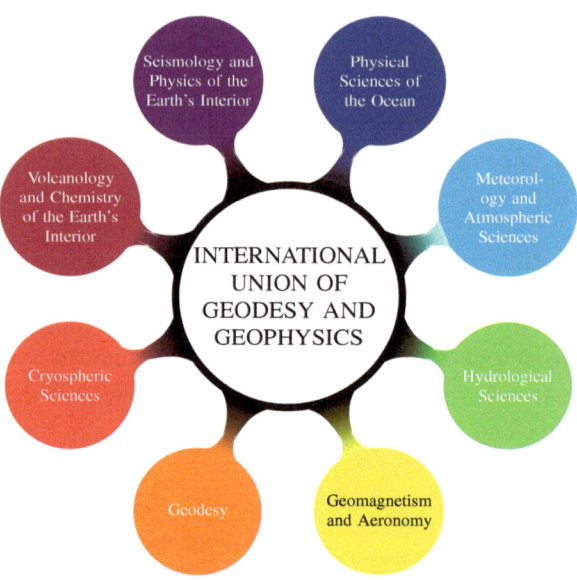

focuses on improving the understanding of the dynamics, chemistry, energetics, and electrodynamics of Earth's atmosphere-ionosphere system. IAGA is one of the eight associations of the *International Union of Geodesy and Geophysics*[3] (IUGG), which is establihed in 1919 and is one of the oldest international geophysics unions. The eight international associations that form the IUGG are presented in Fig. 1.3. Various fields of Earth system science are represented. The Scientific Community on Solar-Terrestrial Physics (SCOSTEP) has recently launched the *Variability of the Sun and Its Terrestrial Impact*[4] (VarSITI) program, in which various coupling processes concerning the middle atmosphere, thermosphere, and ionosphere will be investigated by international researchers within the element of ROSMIC (*Role Of the Sun and the Middle atmosphere/thermosphere/ionosphere In Climate*) (Lübken et al. 2014; Ward et al. 2014). The overarching goal of the ROSMIC project is a better understanding of the impact of the solar activity on the whole atmosphere system. As part of the ROSMIC element, a number of international working groups have been established. ROSMIC's *Coupling by Dynamics* working group specifically focuses on the vertical coupling in the atmosphere-ionosphere system. This working group is currently lead by three scientists from Japan, Germany, and the US.

   The different international organizations around the world increasingly realize the significance of cooperation. For example, the International Commission on the Middle Atmosphere[5] (ICMA), which is one of the IUGG associations, together with SCOSTEP has contributed to the organization of the 5th IAGA International

---

[3]http://www.iugg.org/.

[4]http://www.varsiti.org/.

[5]http://www-mete.kugi.kyoto-u.ac.jp/project/ICMA/.

Workshop on Vertical Coupling in the Atmosphere-Ionosphere System in the summer of 2014 in Turkey. This cooperation has motivated a broad range of scientists to participate in this workshop. For the first time in the framework of an IAGA workshop, atmospheric and solar scientists have met to discuss coupling processes.

1.4 Planetary Science Context

Most research questions of atmospheric and ionospheric physics are studied under great challenging circumstances. One type of challenge I would like to focus on here is of a constructive kind, namely, the rapid increase of observations of the atmosphere-ionosphere system. Most theoretical and modeling studies have to question and improve their underlying assumptions in order to better represent and explain the observed structure of a highly complex atmosphere. The more sophisticated observations become the more in detail the actual properties of the atmospheric system and the coupling processes therein are revealed. The general tendency is an ever increasing complexity in the used tools and applied methodologies. Certainly, the atmosphere-ionosphere system is intrinsically a highly nonlinear system. Therefore, theoretical frameworks and models have to be continuously developed and improved in order to interpret any new discoveries and help plan future observational campaigns.

Complexity should not be interpreted as an obstacle on the way to progress. Quiet the opposite effect actually applies to reality. One advantage of studying physical processes in great detail on Earth is that theoretical and computational models develop quiet rapidly, as mentioned. The availability of detailed observations of the Earth system provides better constraints on the models. Essentially, other planetary systems are subject to the same fundamental laws of physics and, thus, an advancement in Earth sciences often gives rise to, or at least is expected to support, progress in planetary sciences. These two fields in fact should not be separated from each other. Most tools that are developed for Earth's atmosphere can readily be used for other planetary atmospheres. For example, the origin of most general circulation models (GCMs) of other planets are terrestrial. GCMs will be discussed later in more detail in Sect. 6.5.

It is an exciting time for atmospheric and planetary sciences. I could possibly not review all the ongoing international planetary science missions, but I wish to highlight some of them. National Aeronautics and Space Administration's (NASA) Mars Atmosphere Volatile EvolutioN[6] (MAVEN) satellite has successfully reached Mars in September 2014, about 10 months after its launch in November 2013. Its prime mission is to investigate the upper atmosphere and ionosphere of Mars and their interaction with the Sun. In this context, one of the focus topics is the role of the lower atmosphere on atmospheric escape. The Cassini Mission jointly led by NASA, the European Space Agency (ESA), and the Italian space agency have

---

[6]http://lasp.colorado.edu/home/maven/.

so far provided an unprecedented view of Saturn and its moons, such as its largest moon Titan. NASA's New Horizon mission[7] to Pluto and the outskirts of the Solar System made its first encounter with Pluto in January 2015. This challenging flyby mission is expected to provide insight into Pluto's atmospheric processes and shed light into the unexplored parts of the Solar System. As I am writing this text, new data are continuously arriving from various planetary missions and our understanding and imagination of the Solar System (and beyond) are undergoing an every faster transformation.

References

Blanch E, Marsal S, Segarra A, Torta JM, Altadill D, Curto JJ (2013) Space weather effects on earths environment associated to the 2425 October 2011 geomagnetic storm. Space Weather 11:153–168. doi:10.1002/swe.20035

Dehant V, Lammer H, Kulikov Y, Griemeier JM, Breuer D, Verhoeven O, Karatekin Van Hoolst T, Korablev O, Lognonn P (2007) Planetary magnetic dynamo effect on atmospheric protection of early Earth and Mars. Space Sci Rev 129(1–3):279–300. doi:10.1007/s11214-007-9163-9, http://dx.doi.org/10.1007/s11214-007-9163-9

Huang SS (1959) Occurrence of life in the universe. Am Sci 47:397

Lammer H, Bredehft J, Coustenis A, Khodachenko M, Kaltenegger L, Grasset O, Prieur D, Raulin F, Ehrenfreund P, Yamauchi M, Wahlund JE, Griemeier JM, Stangl G, Cockell C, Kulikov Y, Grenfell J, Rauer H (2009) What makes a planet habitable? Astron Astrophys Rev 17(2):181–249. doi:10.1007/s00159-009-0019-z, http://dx.doi.org/10.1007/s00159-009-0019-z

Lübken FJ, Seppälä A, Ward WE (2014) Project ROSMIC. VarSITI Newsl 1:7–9

Ward W, Lübken FJ, Seppälä A (2014) ROSMIC, a new project in the SCOSTEP VarSITI program. In: 40th COSPAR Scientific Assembly, Moscow, Russia

---

[7]http://pluto.jhuapl.edu/.

Chapter 2
Introduction to Atmospheric Physics

Fundamental Concepts

The book of nature is written in the language of mathematics.
—Galileo Galilei (15–16th century A.D.)

Abstract A brief introduction to the physics of the neutral atmospheres is given by reviewing some fundamental concepts. Some fluid mechanical principles of atmospheric physics are highlighted. To a good approximation, Earth's atmosphere and most other planetary atmospheres can be treated as an ideal gas that is in hydrostatic equilibrium. Thermodynamical laws are powerful in describing the basics of energy conservation in the atmosphere. Transport phenomena, such as diffusion, conduction, viscosity, and radiation are discussed briefly in order to provide some insight in the mechanisms of energy and momentum transfer processes in planetary atmospheres.

Keywords Continuum hypothesis · Geopotential · Potential temperature · Transport processes · Thermodynamic laws · Hydrostatic equilibrium

2.1 Introduction to Earth's Atmosphere

Earth's atmosphere, commonly known as "air" in daily life, is a thin layer of gas that surrounds the lithosphere, i.e., the upper layer of solid Earth. The atmosphere is pervaded by a complex geomagnetic field that is produced internally within Earth's core by an active magnetohydrodynamic (MHD) dynamo. This intrinsic geomagnetic field extends into the outer space by several Earth radii and largely shields Earth from high-energetic particles from the Sun and other astronomical sources, such as, the galactic cosmic rays. Existence of air with a significant amount of oxygen content with the right pressure permits the existence of life on our planet Earth. The

© The Author(s) 2015   9
E. Yiğit, *Atmospheric and Space Sciences: Neutral Atmospheres*,
SpringerBriefs in Earth Sciences, DOI 10.1007/978-3-319-21581-5_2

Fig. 2.1 Vertical
atmospheric layers:
Troposphere, Stratosphere,
Mesosphere, Thermosphere,
which coexists with the
ionosphere. Not to scale

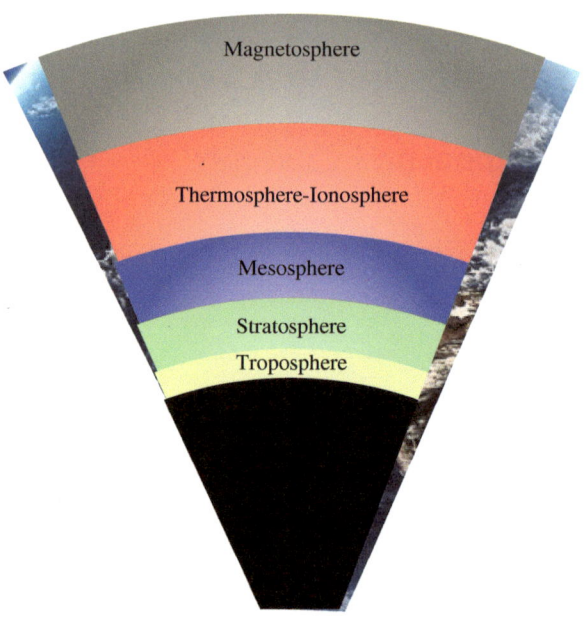

atmosphere also protects life on the surface from Sun's dangerous radiation. The
study of the structure and composition of planetary atmospheres is of great relevance
to the subject of habitability.

Figure 2.1 demonstrates a simple illustration of the vertical layers of Earth's
atmosphere. The lower portion of the atmosphere is called the *troposphere* and contains approximately 90 % of the total atmospheric mass and is the region where the
meteorological processes (i.e., weather) occur. These processes affect the day-to-
day human life. With increasing altitude, the density drops off rapidly, causing the
increase in the intensity of dynamical interaction processes occurring at higher altitudes. This is to say that a given amount of energy deposition would produce larger
effects, e.g., heating, in the upper atmosphere than at lower altitudes.

Compared to the mean radius of Earth ($r_e \approx 6378$ km), the atmosphere is relatively
thin. Its overall structure is controlled by a complex interplay of chemical and physical processes that couple the different atmospheric layers downward and upward
over a large range of altitudes. If only in-situ processes are considered, the lower
atmosphere (troposphere) is dominated primarily by meteorological processes while
the upper portion of the atmosphere (*thermosphere-ionosphere*) is subject to electromagnetic and electrodynamical processes of solar and geomagnetic origin. The middle atmosphere (*stratosphere-mesosphere*) is the interface region between the lower
and upper atmosphere that is influenced from below and above. The ozone layer, the
protective shield from Sun's EUV radiation, is situated within the stratosphere. The
physical structure of this layer has influence on the penetration of radiation and thus
has importance for human health. Small- and large-scale waves that are produced by

Table 2.1 Some planetary parameters for Mercury, Venus, Earth, and Mars

	Radius [km]	Mean sun-planet distance [AU]	Rotational period [hr]
Mercury	2436	0.387	1403.3
Venus	6052	0.723	5832.2
Earth	6378	1	23.93
Mars	3396	1.523	24.6

$1\,\mathrm{AU} \approx 1.496 \times 10^8$ km

various sources shape the circulation of the middle atmosphere. The importance of the lower atmosphere in influencing the upper atmosphere across the middle atmosphere is being increasingly acknowledged by the research community, following the recent progress in modeling and observations of vertical coupling processes.

Depending primarily on the solar and geomagnetic activity, the atmosphere extends from the surface up to 500–900 km. In particular, the 11-year solar activity cycle and geomagnetic processes, such as magnetic storms associated with solar flares and coronal mass ejections (CMEs), are dominant geophysical processes that largely control the overall morphology of the upper layers of the atmosphere, i.e., thermosphere-ionosphere. The vertical structure of the atmosphere-ionosphere system will be presented in more detail in Chap. 4.

The essential planetary parameters of Earth are described in Table 2.1 along with their values for the innermost planet Mercury, and the two with respect to Earth neighboring terrestrial-like planets, Mars and Venus. It is noteworthy that Earth has a rotational period similar to Mars and a planetary radius similar to Venus. With a mean distance of about 0.3 AU Earth is closer to Venus than to Mars at a mean distance of 0.5 AU, where 1 AU is approximately 149.6 million kilometers. Mercury's radius is about 2.5 times smaller than Earth's.

In this chapter, my goal is to present a brief overview of some fundamental concepts of atmospheric dynamics and thermodynamics that are useful in the description of planetary atmospheres. Later chapters will survey governing equations of atmospheric physics (Chap. 3), the geospace environment and the vertical structure of Earth's atmosphere (Chap. 4), atmospheric waves and their effects (Chap. 5), and the general circulation of the atmosphere (Chap. 6). In particular, Chaps. 5 and 6 will include state-of-the-art research topics in atmospheric waves and dynamics.

2.2 Basic Atmospheric Parameters and Coordinate System

The atmosphere is a thermodynamic fluid and can therefore be described by a set of thermodynamic parameters. Often the expressions of "parameter" and "variable" are used interchangeably. Some fundamental atmospheric parameters are temperature T, pressure p, and volume V. Physical variables, such as, temperature, mass density

Fig. 2.2 Illustration of the
Cartesian and spherical
geometries, where x, y, and
z are the rectangular
coordinates; and r, ϕ, and θ
are the radial distance,
azimuthal angle, and latitude,
respectively. Latitude and
longitude are in radians

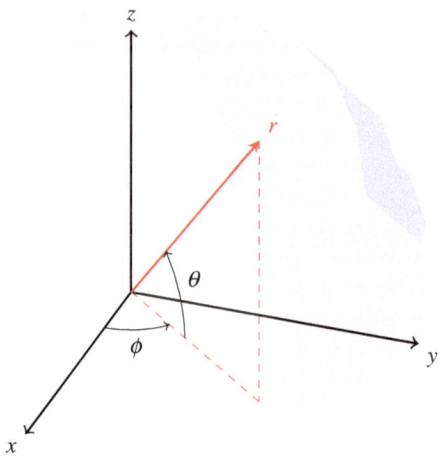

ρ, pressure, and humidity, express the physical state of the atmosphere. *Kinematic variables*, such as wind velocity $\mathbf{u} = (u, v, w)$ with zonal (u, East-West), meridional (v, North-South) and vertical (w, upward-downward) wind components, describe the motion of air particles or systems of air particles. *Direct parameters*, such as, p, T, surface wind, and relative humidity, can be directly measured; parameters, such as, momentum $\mathbf{p}$, ρ, and potential temperature θ_T (see Sect. 2.11) are evaluated using equations of physics and are thus called derived parameters.

In a realistic atmosphere, all properties are time-dependent and vary in three-dimensions represented by the position vector $\mathbf{r}$. Time is denoted by the variable t, while the space variables depend on the choice of the coordinate system. In the rectangular coordinate system or the Cartesian[1] coordinate system, we have $\mathbf{r} = (x, y, z) = x\,\hat{\mathbf{i}} + y\,\hat{\mathbf{j}} + z\,\hat{\mathbf{k}}$, where $\hat{\mathbf{i}}, \hat{\mathbf{j}}$, and $\hat{\mathbf{k}}$ are the associated unit vectors. Sometimes unit vectors are denoted by $\hat{\mathbf{e}}$, for example, as $\hat{\mathbf{e}}_x$, $\hat{\mathbf{e}}_y$, and $\hat{\mathbf{e}}_z$. In spherical polar geometry, $\mathbf{r} = (r, \phi, \theta) = r\,\hat{\mathbf{r}} + \phi\,\hat{\boldsymbol{\phi}} + \theta\,\hat{\boldsymbol{\theta}}$, where r is the radial distance, ϕ is the azimuthal angle, and θ is the latitude, as illustrated in Fig. 2.2.

The amount of a given thermodynamic quantity per unit time (second) and per area element m^{-2} is understood as the flux of the quantity through a given area. For example, in atmospheric physics, one often speaks of momentum, energy, mass, and particle flux that provide important diagnostic information.

2.3 Continuum Hypothesis

Fluid mechanics is concerned with the behaviour of matter on a macroscopic scale, i.e., a scale that is much larger than the distances between individual molecules. Microscopic scale is related to atomic scales. Very often the molecular structure of

---

[1]This coordinate system is called in honor of the mathematician René Descartes.

Fig. 2.3 A volume element $\delta V = \delta x\,\delta y\,\delta z$ in Cartesian coordinates

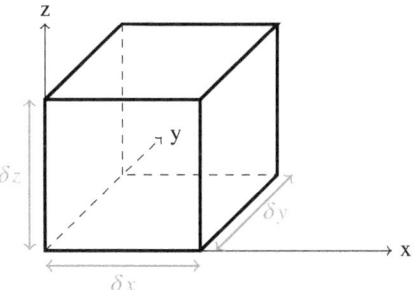

matter does not need to be taken into account. This assumption is justified provided that the flow length scales are much larger than the molecular mean free path. An atmospheric air parcel is typically represented by a volume element δV, which is assumed to be infinitesimally small and contain many atoms and molecules. The word "hypothesis" means originally a basis or a basis of an argument. In a scientific context, it is a tentative explanation of a phenomenon. A "continuum" implies continuity. Therefore, the "continuum hypothesis" states that the macroscopic characteristics of a fluid matter containing many molecules is the same and perfectly continuous in structure. The momentum and energy of particles contained in a given air parcel are then uniformly distributed within the volume element. Such an infinitesimally small volume element is often imagined simply in terms of the Cartesian coordinates as $\delta V = \delta x\,\delta y\,\delta z$ illustrated in Fig. 2.3. This simplification is very convenient in deriving conservation laws of physics.

2.4 Atmospheric Composition

The atmosphere, a gaseous matter, is composed of atoms and molecules. An atom is made up of electron(s) and a nucleus that includes the proton(s) and the neutron(s). Complex organization of these subatomic particles in the presence of various fundamental forces gives rise to a spectrum of atoms listed in the periodic table. The lightest atom is the hydrogen atom, $_{1}^{1}H$, composed of one electron and one proton, where the proton mass is $m_p = 1.672 \times 10^{-27}$ kg and the electron mass is $m_e = 9.109 \times 10^{-31}$ kg, and thus $m_p \gg m_e$. While the strong force keeps the nucleus together, the electromagnetic force is responsible for interactions between charged particles. Molecules are composed of two or more atoms.

Earth's lower and middle atmosphere is mixed by turbulence. Table 2.2 lists the major and minor atmospheric constituents according to their abundance from left to right. About 78 % of the atmosphere consists of Nitrogen (N_2), while only about 21 % is the molecular Oxygen (O_2). Minor constituents such as Carbon Dioxide (CO_2), Argon (Ar), and water vapor (H_2O) together make up about 1 %. 0.03 % of CO_2 fractional composition is sometimes expressed in terms of

Table 2.2 Fractional composition of the terrestrial atmosphere in terms of volume mixing ratios in percentage

O_2	N_2	CO_2	Ar	H_2O
21	78	0.03	0.9	0–2

particle per million (ppm) and corresponds to 300 ppm. In addition to the water vapor, CO_2 molecule is an important greenhouse gas. The variation of its concentration has wide-reaching impact on Earth's climate and is a central topic of research in climate sciences.

The atmosphere including the troposphere, stratosphere, and mesosphere, where the neutral species are well-mixed by turbulence, is often referred to as the *homosphere*. Above the homosphere in the thermosphere, diffusive separation gradually overtakes and the concentrations of the different species drop off differently with increasing altitude, depending on their weights. In the upper atmosphere, neutral mass density can demonstrate significant geographical and temporal variations.

2.5 Universal Gravity

The law of universal gravitation introduced by Isaac Newton in 17th cc states that every matter in the universe attracts every other matter with a force F_{12} that is proportional to the product of their masses m_1 and m_2 and inversely proportional to the square of the distance r between them. In other words, "everything pulls everything else". This law is given mathematically by

$$F_{12} \propto \frac{m_1 m_2}{r^2}, \tag{2.1}$$

where

$$F_{12} = -F_{21}, \tag{2.2}$$

meaning that the force exerted by m_1 on m_2, F_{12}, and the force exerted my m_2 on m_1, F_{21}, have the same magnitude but opposite directions. The constant of proportionality is the G,

$$F_{12} = -G \frac{m_1 m_2}{r^2}. \tag{2.3}$$

The numerical value of G can be determined experimentally by measuring the force between two test masses with a known separation distance. Although for small bodies the gravitational force is very small, the Cavendish balance can be used to determine it. With this method the constant is found to be $G = 6.67 \times 10^{-11}$ N m^2 kg^{-2}. The value of G has great significance for the formation of the universe after the Big Bang. A slightly smaller value could have meant that there could not have been enough gravitational force between matter to clump together to form more complex matter. A stronger G could have implied that the expansion of the universe could have not taken place.

In terms of fundamental particles, you can consider the gravitational attraction between a proton and an electron, which amounts to $\sim 10^{-47}$ N at an atomic distance of 1 Å. In planetary atmospheres, the most obvious effect of gravity is that everthing is accelerated toward the center of the planet and it is a body force in the equation of motion. The associated acceleration of the mass m in a planet of mass M due to the force F_{Mm} exerted by the planet on m is

$$g = \frac{F_{Mm}}{m} = G\frac{M}{r^2}, \tag{2.4}$$

where the distance $r = r_p + z$ is the sum of planetary radius and the altitude from the surface. For Earth, with $M_e = 5.97 \times 10^{24}$ kg the mean gravitational acceleration on the surface is $g_0 \approx 9.80\,\text{m s}^{-2}$. For Mars, with $M_m = 0.1\,M_e$ and $r_m = r_e/2$ the mean surface gravitation acceleration is $3.92\,\text{m s}^{-2}$, which is 2.5 smaller than on Earth. The altitude variation of the gravitational acceleration is approximately given by

$$g(z) = \frac{g_0}{1 + \frac{z}{r_E}}, \tag{2.5}$$

For $z \ll r_E$ we have $g \approx g_0$.

2.6 Equation of State: Ideal Gas Law

A relation between the fundamental thermodynamic variables, such as, pressure p, temperature T, volume V, and number of particles (molecules) N in a thermodynamic system is given by an equation of state f as

$$f(p, T, V, N) = 0. \tag{2.6}$$

This property states that only a certain number of the properties of a substance can be given arbitrary values in a thermodynamic system. The specific form of an equation of state depends on the substance.

An ideal gas is defined in an atomistic view as a collection of gas of any species in which no forces operate between the individual gas particles. It has been experimentally found that there is a distinct relationship between the pressure and the temperature of an ideal gas. Namely, pressure times molal specific volume v (i.e., volume per mole V/n) divided by temperature is equal to a constant R

$$\frac{pv}{T} = R, \tag{2.7}$$

where the universal gas constant R is

$$R = 8.317 \text{ J mole}^{-1}\text{K}^{-1}. \tag{2.8}$$

Because there are a large number of atoms in a given volume, chemists have introduced the concept of "mole" in order to count more easily. One mole contains
$N_A = 6.02 \times 10^{23}$ particles, where N_A is known as Avogadro's number[2] and is
defined by the number of carbon-12 atoms contained exactly in 12 g of carbon-12
($^{12}_{6}$C). So the amount of matter in a volume element can be measured in terms of
moles as well. For example, a mole of oxygen $^{16}_{8}$O is 16 g. The number of atoms in
a mole times the Boltzmann constant yields the universal gas constant R

$$R = N_A k_b, \tag{2.9}$$

where the Boltzmann constant $k_b = 1.381 \times 10^{-23}$ J K^{-1}. In turn, at a fixed temperature and volume, the total number of atoms can be determined.

The ideal gas law in terms of the volume and the number of moles is then

$$pV = nRT. \tag{2.10}$$

Equation (2.10) can be expressed in terms of the density ρ, which is of interest in
planetary atmospheres. With $V = m/\rho$ combined with an expression for the total
mass m in terms of the mean molecular mass m_m as $n = m/m_m$, we can express
volume

$$V = \frac{nm_m}{\rho}, \tag{2.11}$$

and substitute in (2.10), which yields

$$p = \rho \frac{R}{m_m} T. \tag{2.12}$$

The mean molecular mass varies with altitude in planetary atmospheres. On Earth,
$m_m \approx 28.8$ g mol^{-1} in the lower and middle atmosphere and decreases in the
thermosphere. For practical calculations a typical value of m_m can be assumed to
define

$$R_* = \frac{R}{m_m} \approx 288.7 \text{ J kg}^{-1} \text{ K}^{-1}, \tag{2.13}$$

yielding a simplified equation of state for an ideal gas

$$p = \rho R_* T. \tag{2.14}$$

---

[2]Named after the Italian physicist Amedeo Avogadro.

The ideal gas law is a limiting case for the behaviour of a real gas, which is described by the Van der Vaals law. For planetary atmospheres, the ideal gas law is a very good approximation.

At constant temperature, pressure is inversely proportional to volume

$$p \propto V^{-1}, \tag{2.15}$$

called the Boyle-Mariotte law. At constant volume, pressure is proportional to temperature

$$p \propto T, \tag{2.16}$$

referred to as the Guy-Lussac law.

2.7 Thermodynamic Laws

Thermodynamics is the branch of physics that studies the relationship among the various properties of matter, without the knowledge of internal structure of the matter. The *laws of thermodynamics* contain qualitatively all the principles of thermodynamics. The first law of thermodynamics states that the conservation of energy given by

$$\Delta U = \Delta Q + \Delta W, \tag{2.17}$$

where ΔU, ΔQ, and ΔW are the changes in internal energy, heat flow, and in work done. This law states that the total change in the internal energy of a system is given by the sums of changes in heat flow and the work done. The differential form of (2.17) is then

$$dU = dQ + dW, \tag{2.18}$$

where $dW = -pdV$ is the work done on the system.

The second law of thermodynamics originally hypothesized by Carnot implies that heat on its own cannot flow from a cold to a warm object. This is the concept that is familiar to us from the day-to-day life. In a hot summer day, if you leave the balcony door open, then your room that had been cooled by an air conditioner will warm up by heat flow from outside into the room. Specifically, the entropy (disorder) in an isolated system either increases or remains constant.

A thermodynamical system can be taken from one state to another state by various thermal and/or dynamical processes. This change can take place in different ways. For example, in adiabatic processes, the temperature of a system is independent of its environment and the kinetic and potential energy of the system do not change.

2.8 Heat Capacity

Heat capacity C is an important concept in atmospheric thermodynamics. It expresses the amount of energy required to raise the temperature of a substance by 1 K. In general, we have

$$C \equiv \frac{\mathrm{d}Q}{\mathrm{d}T}. \qquad \left[\frac{\mathrm{J}}{\mathrm{K}}\right] \tag{2.19}$$

The first law (2.18) can be written in terms of the specific quantities

$$\mathrm{d}\varepsilon = \mathrm{d}q_s - p\mathrm{d}v, \tag{2.20}$$

where ε is the internal energy density (that is, internal energy per unit mass), q_s is the heat per unit mass, and $v = \rho^{-1}$ is the specific volume. Additionally, let us expand the change in the internal energy density $\varepsilon = \varepsilon(T, v)$ in terms of the changes in temperature and specific volume

$$\mathrm{d}\varepsilon = \left(\frac{\partial \varepsilon}{\partial T}\right)_v \mathrm{d}T + \left(\frac{\partial \varepsilon}{\partial v}\right)_T \mathrm{d}v, \tag{2.21}$$

where the subscripts v and T indicate that the partial derivatives are taken at constant specific volume and constant temperature, respectively. Inserting (2.21) into (2.20) yields

$$\mathrm{d}q_s = \left(\frac{\partial \varepsilon}{\partial T}\right)_v \mathrm{d}T + \left[\left(\frac{\partial \varepsilon}{\partial v}\right)_T + p\right]\mathrm{d}v, \tag{2.22}$$

In general, heat flow can occur at constant volume or constant pressure. In the case of constant volume ($\mathrm{d}v = 0$), we get the specific heat capacity at constant volume c_v

$$\mathrm{d}q_s = \left(\frac{\partial \varepsilon}{\partial T}\right)_v \mathrm{d}T \iff \mathrm{d}q_s = c_v \mathrm{d}T. \tag{2.23}$$

At constant pressure ($\mathrm{d}p = 0$), one has

$$\mathrm{d}q_s = c_p \mathrm{d}T. \tag{2.24}$$

Note that the specific heat capacity has the unit of J kg^{-1} K^{-1}. On Earth, for dry air, $c_p = 1004$ J kg^{-1} K^{-1} and $c_v = 717$ J kg^{-1} K^{-1}. In comparison, water has a specific capacity of $c_p = 4180$ J kg^{-1} K^{-1}. The ratio of the specific heat capacities is given by

$$\gamma \equiv \frac{c_p}{c_v}, \tag{2.25}$$

where γ is sometimes called the specific heat ratio. For monoatomic gases, such as, He and Ar $\gamma = 1.67$ and for diatomic gases, e.g., for N_2 and O_2, $\gamma = 1.4$.

2.9 Geopotential

Consider a volume of air at rest on the surface of Earth. Then, the fluid volume is subject to two types of forces: (i) pressure gradient force due to the exponentially decreasing pressure with altitude; (ii) Coriolis forces due to the rotation of Earth plus the gravitational force. The conservative body force that is associated with a potential Φ is called the geopotential. It is composed of Earth's gravitational potential and the centrifugal potential related to Earth's rotation. Geopotential is the amount of energy required to move a unit mass vertically upward from the surface to a reference level z.

$$\Phi(z) = \int_0^z g(z')\, dz' \quad [\text{m}^2\,\text{s}^{-2}] \tag{2.26}$$

If the sea on Earth were at rest then the geopotential surface would coincide with the sea surface, which is taken as a reference level. For most calculations, the gravitational acceleration g is assumed to be constant as $g_0 = 9.8\ \text{m s}^{-2}$. Then,

$$\Phi(z) \approx g_0 z, \tag{2.27}$$

where the vertical coordinate z denotes the vertical distance from the reference level and is typically called the geopotential height,

$$Z = \frac{\Phi}{g_0}. \tag{2.28}$$

So, to move a mass of 1 kg over 1 m vertical distance in the gravitation field of Earth, one would need $\sim$9.8 J.

2.10 Hydrostatic Equilibrium

Under the effect of gravity, the atmospheric pressure p decreases exponentially with increasing height. In order to obtain an expression for the vertical variation of pressure, consider a column of air with altitude extent dz and bottom surface area of A. Then the pressure exerted by the air inside the column is given by the force divided by the area

$$dp = \frac{dF}{A}. \tag{2.29}$$

The infinitesimal force is $dF = -g\,dm = -g\rho\,dV$, with the volume element $dV = A\,dz$. The pressure dp becomes

$$dp = \frac{dF}{A} = -\frac{g\,\rho\,A\,dz}{A}. \tag{2.30}$$

The hydrostatic equilibrium (or balance) is then given by

$$\frac{dp}{dz} = -g\rho. \tag{2.31}$$

Equation (2.31) expresses that the downward gravitational acceleration is balanced by the upward directed acceleration associated with the negative upward pressure gradient. Without Earth's gravity, air would be accelerated into space. Hydrostatic acceleration implies that the vertical acceleration of air is negligible, though vertical air speed can be non-zero. In processes that may involve nonhydrostatic effects (e.g., Yiğit and Ridley 2011) the hydrostatic condition (2.31) does not apply.

From the hydrostatic relation, the vertical variation of pressure can be derived. Applying separation of variables in (2.31), representing the density with the ideal gas law (2.12) and then writing with the left-hand side and right-hand side with integrals with limits from some reference levels p_0 to p and z_0 to z, respectively, give

$$\int_{p_0}^{p} \frac{dp}{p} = -\int_{z_0}^{z} \frac{m(z)\,g(z)}{R\,T(z)}\,dz, \tag{2.32}$$

and finally integrating the left-hand side with respect to pressure yields

$$p(z) = p_0 \exp\left(-\int_{z_0}^{z} \frac{mg}{RT}\,dz\right), \tag{2.33}$$

in terms of the scale height H

$$H \equiv \frac{RT}{mg}, \tag{2.34}$$

we get

$$p(z) = p_0 \exp\left(-\int_{z_0}^{z} \frac{1}{H(z)}\,dz\right). \tag{2.35}$$

Over small vertical distances, the scale height may vary slowly in the lower atmosphere and can be assumed to be constant for practical purposes. Equation (2.35) can be approximated

$$p(z) = p_0 \exp\left(-\frac{z - z_0}{H}\right), \tag{2.36}$$

which shows that the pressure falls off exponentially with height. In the upper atmosphere, because of diffusive separation, one speaks of the individual scale heights of the different species. For a neutral species i with molecular mass m_i, we then have

$$H_i(z) = \frac{R\,T(z)}{m_i\,g(z)} \tag{2.37}$$

Overall, the scale height can vary significantly with height because of the large variations of temperature and the mean molecular mass with height. The concept of scale height also applies to ionized species.

2.11 Potential Temperature

The potential temperature is the temperature a parcel of air at pressure p and temperature T would obtain if it were expanded or compressed adiabatically to a standard pressure $p_s = 1000$ hPa.

$$\theta_T = T\left(\frac{p_s}{p}\right)^{R/c_p}, \tag{2.38}$$

where c_p is the specific heat at constant pressure and T is temperature. Equation (2.38) is known as Poisson equation as well. For dry adiabatic conditions, the potential temperature is conserved. Sometimes, instead of temperature and pressure, potential temperature and pressure are used as state variables. Also, the stability of the atmosphere is measured in terms of the potential temperature gradient.

2.12 Atmospheric Stability

The variation of neutral temperature with altitude in a planetary atmosphere is an important criterion that shapes the stability of the atmosphere.

2.12.1 Lapse Rate

Temperature lapse rate Γ describes the rate of temperature decrease with increasing altitude:

$$\Gamma = -\frac{\partial T}{\partial z} \tag{2.39}$$

Combining the Poisson equation (2.38), the ideal gas law and the law of hydrostatic equilibrium (2.31) yields

$$\frac{T}{\theta}\frac{\partial\theta_T}{\partial z} = \frac{\partial T}{\partial z} + \frac{g}{c_p} \qquad (2.40)$$

Dry adiabatic lapse rate Γ_d is given when $\frac{\partial\theta_T}{\partial z} = 0$

$$-\frac{\partial T}{\partial z} = \frac{g}{c_p} \equiv \Gamma_d \qquad (2.41)$$

2.12.2   Static Stability

In a realistic atmosphere the potential temperature varies with altitude. The actual lapse rate Γ would differ from the dry adiabatic lapse rate Γ_d. This difference is obtained by combining Eqs. (2.40)–(2.41):

$$\frac{T}{\theta}\frac{\partial\theta}{\partial z} = \Gamma_d - \Gamma \qquad (2.42)$$

A parcel of air that undergoes adiabatic displacements can be either positively or negatively buoyant and tends to return to its equilibrium position. If the actual lapse rate is smaller than the dry adiabatic lapse rate, i.e., $\Gamma < \Gamma_d$, then the potential temperature increases with altitude ($\frac{\partial\theta}{\partial z} > 0$). The air parcel that is displaced adiabatically from its equilibrium position will be positively buoyant when displaced downward and negatively buoyant when displaced upward. Under these conditions Earth's atmosphere is stably stratified. Adiabatic oscillations of a fluid parcel around its equilibrium in a stably stratified atmosphere are called buoyancy oscillations. The frequency of these oscillations is given by the Brunt-Väisälä frequency N. In stably stratified atmospheres, gravity waves can propagate freely. If the atmospheric lapse rate exceeds the dry adiabatic lapse rate, i.e., $\Gamma > \Gamma_d$, it implies that ($\frac{\partial\theta}{\partial z} < 0$) and the atmosphere can become convectively unstable.

2.13 Transport Phenomena

Any non-equilibrium condition in a system involves some type of transport process. Basic transport processes within fluids are *diffusion*, *viscosity*, and thermal *conduction*. All transport processes have an associated transport coefficient that quantifies the degree of the transport. The coefficients of diffusion D, viscosity μ, and conduction κ are summarized along with their dimensions in Table 2.3.

Table 2.3 Transport processes and the associated transport coefficients

Transport process	Coefficient	Dimension
Diffusion	D	$m^2 \, s^{-1}$
Viscosity	μ	$kg \, m^{-1} \, s^{-1}$
Conduction	κ	$J \, s^{-1} \, m^{-1} \, K^{-1}$

D is the diffusion coefficient, μ is the coefficient of viscosity, and κ is the thermal conductivity coefficient

2.13.1 Diffusion

Diffusion involves molecular motion. It is a consequence of frequent, stochastically distributed collisions between particles. Imagine a container of a background gas in thermal equilibrium. We then introduce a small amount of a different gas in a random place in the container. The particles of the new gas then spread gradually through the background gas, colliding with the particles of the background gas. This new gas is said to "diffuse" through the background gas and the process is called diffusion.

In a system of inhomogeously distributed particles, Fick's law expresses the net particle flux Φ, which is proportional to the negative gradient of the particle number density. In $1-D$ it is given by

$$\Phi = -D\frac{dn}{dx},\tag{2.43}$$

where D is the diffusion coefficient and n is the number density, that is, number of particles per unit volume, of particles. If we consider three-dimensional diffusion, then the most generic form of particle flux is

$$\Phi = -\mathbf{D} \cdot \nabla n,\tag{2.44}$$

where $\mathbf{D}$ is the anisotropic diffusion tensor and the Φ is the particle flux vector. Assuming $n = n(x, t)$, particle flux can be expressed considering the continuity (Sect. 3.7) requirement

$$\frac{\partial n}{\partial t} + \frac{\partial (nu)}{\partial x} = 0,\tag{2.45}$$

where u is the speed in the positive x-direction and nu is the particle flux in the x-direction. Combination of (2.43) with (2.45) yields the diffusion equation.

$$\frac{\partial n}{\partial t} = D\frac{\partial^2 n}{\partial x^2},\tag{2.46}$$

assuming isotropic homogeneous diffusion. In three dimensions one gets

$$\frac{\partial n}{\partial t} = D\,\nabla^2 n, \tag{2.47}$$

where $\nabla^2 = \frac{\partial^2}{\partial x^2} + \frac{\partial^2}{\partial y^2} + \frac{\partial^2}{\partial z^2}$ is the Laplace operator in Cartesian coordinates.

In processes that involve advective processes as well, the diffusion equation can be extended to the advective-diffusion equations. More on the formalism of advection and diffusion can be found in the work by Medvedev and Greatbatch (2004).

2.13.2 Viscosity

Viscosity can be considered as a frictional force. If there is a velocity shear (gradient) perpendicular to the flow in a fluid, then transport of momentum occurs within the flow in order to balance the velocity gradient. Therefore, in any nonuniform fluid flow, we expect viscosity to be present. The flow is said to possess viscosity because of the thermal motion of particles perpendicular to the flow direction.

Let us imagine a one-dimensional shear flow directed in the positive x-direction (zonal direction) given by the speed u, that is, $\mathbf{u} = (u(y), 0, 0)$. Let the flow possess an increasing gradient of velocity in the y-direction (meridional direction), as illustrated in Fig. 2.4. Then neigbouring fluid layers exert some stress, that is, force per unit contact area, on each other. This configuration leads to the transport of y-momentum in x-direction denoted by τ_{xy} and is experimentally found to be proportional to the gradient of the x-component of the flow in the y-direction

$$\tau_{xy} = -\mu\frac{\partial u}{\partial y}, \tag{2.48}$$

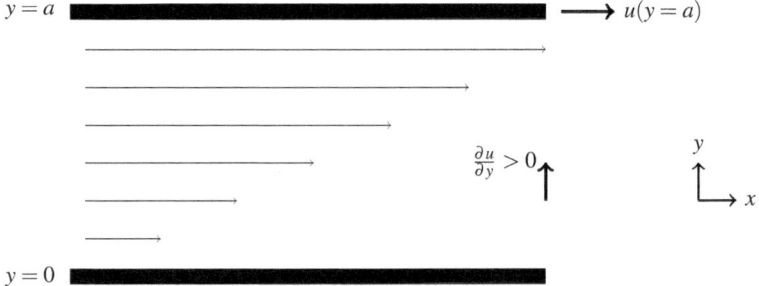

Fig. 2.4 Illustration of viscosity in the x–y plane with u denoting the x-component of the velocity. *Thick lines* denote the *top* and the *bottom* boundaries, while the *thin arrow* show the velocity components in away from the boundaries, shown here declining in the negative y-direction and increasing in the positive y-direction.l

where the proportionality factor μ is the coefficient of viscosity, which is given in terms of the kinematic viscosity $\mu = \nu\rho$. In the absence of velocity shear there is no induced stress because of viscosity. In other words, the stress arising from the presence of viscosity acting perpendicular to the flow is called *shear stress*. As the shear stress is proportional to the velocity gradient, the fluid said to be *Newtonian*. The physical significance of shear stresses in a fluid arises from the fact that any imbalance in stress can lead to a net body force on the fluid. In our example, τ_{xy} expresses the transfer of y-momentum in the x-direction, resulting from a velocity shear in the y-direction. In a three-dimensional case, additional contribution to momentum flux in x-direction can come from a velocity shear in the z-direction, that is, $\tau_{xz} > 0$ in the presence of shear in the z-direction. In general, the force per unit mass in the i-th direction resulting from viscous processes is

$$a_i = \frac{F_i}{m} = \frac{1}{\rho}\frac{\partial \tau_{ij}}{\partial x_j} = -\frac{1}{\rho}\frac{\partial}{\partial x_j}\left(\mu\frac{\partial u_i}{\partial x_j}\right), \tag{2.49}$$

where τ_{ij} is the flux of x_j-component of momentum in the x_i-direction. It is important to note that the total stress σ_{ij} arising in a fluid is composed of pressure (normal stress) and shear stress τ_{ij}

$$\sigma_{ij} = -p\delta_{ij} + \tau_{ij}, \tag{2.50}$$

where δ_{ij} is the Kronecker delta with the property

$$\delta_{ij} = \begin{cases} 1 \text{ if } i = j \\ 0 \text{ if } i \neq j \end{cases} \tag{2.51}$$

Viscous force arising from the relative motion of different fluid layers is always a frictional force that counteracts the velocity variations in the flow.

2.13.3 Conduction

Conduction describes the flow of thermal energy per unit area per unit time, i.e., heat flux. In response to a temperature gradient, heat is conducted. Heat flows in the direction of decreasing temperature, that is, from hot to cold regions in accordance with the second law of thermodynamics. In a uniform temperature distribution, heat conduction would be zero. Fourier's law describes the heat conduction in a stationary medium

$$q = -\kappa\frac{\partial T}{\partial x}, \qquad [\text{J s}^{-1}\,\text{m}^{-2}] \tag{2.52}$$

where q is heat flux, κ is the thermal conductivity, and T is temperature. Heat flux describes the transfer of heat in terms of energy per unit area per unit time. The negative sign is because the heat flow is in the opposite direction to the temperature gradient. Fourier law of heat conduction assumes a steady-state heat conduction, meaning that the temperature distribution in the system is independent of time.

If we assume that the temperature distribution in a closed system is changing with time, for example, due to time-varying boundary conditions, then we speak of an unsteady (time-dependent) heat conduction process. The change of the internal energy per unit mass ε is then

$$-\frac{\partial \varepsilon}{\partial t} = \frac{1}{\rho}\frac{\partial q}{\partial x} = -\frac{\partial}{\partial x}\left(\kappa \frac{\partial T}{\partial x}\right). \tag{2.53}$$

With $\varepsilon = c_v T$ and assuming constant thermal conductivity, we get the *unsteady heat conduction equation*

$$\frac{\partial T}{\partial t} = \alpha \frac{\partial^2 T}{\partial x^2}. \tag{2.54}$$

In three dimensions, we have

$$\frac{\partial T}{\partial t} = \alpha \, \nabla^2 T, \tag{2.55}$$

where $\alpha = \kappa \, (\rho \, c_v)^{-1}$ is the thermal diffusivity.

2.13.4 Radiation

Radiation is transport of energy (heat) by means of electromagnetic wave propagation. Sun's energy reaches us via radiation. Every matter in nature emits energy in form of radiation. The characteristics of the emission depends on the temperature of the body. At higher temperatures higher frequency (shorter wavelength) emission takes place. Specifically, the rate of emission (energy radiation) associated with a body is proportional to the surface area A and to the fourth power of temperature T^4. The Stefan-Boltzmann law is given by

$$\frac{\mathrm{d}Q}{\mathrm{d}t} = A \, \varepsilon_r \, \sigma_b \, T^4, \qquad [\mathrm{W}] \tag{2.56}$$

where $\mathrm{d}Q/\mathrm{d}t$ is the heat flow, the proportionality constants ε_r is the radiative emissivity and $\sigma_b = 5.67 \times 10^{-8}$ W m^{-2} K^{-4} is the Stefan-Boltzmann constant. The emissivity varies between 0 and 1 and is determined with respect to an ideal radiating body with a given surface and at a given temperature. The net heat flow depends on the temperature difference between the emitting body and its surroundings. For example, the human body is about $10\,^{\circ}$C warmer than the average room temperature. An emissivity of unity and body area of 1 m^2 yield a net heat flow of about 60 W from the human body.

The radiation field is described in terms of the amount of radiant energy dE_ν in a given frequency interval $(\nu, \nu + d\nu)$, which is transported across an area element dA in the direction confined to an element of the solid angle $d\omega$, during an time interval of dt (Chandrasekhar 1960). In terms of the intensity I_ν associated with the frequency ν, the infinitesimal radiant energy is expressed as

$$dE_\nu = I_\nu \cos \alpha \, dA \, d\omega \, d\nu \, dt. \tag{2.57}$$

Integration over all frequency bands yields the total radiant energy.

2.14 Richardson Number

The Richardson number R_i is a measure of evolution of turbulence in the atmosphere and describes atmospheric stability. It is given by

$$R_i = \frac{N^2}{\left(\frac{\partial u}{\partial z}\right)^2 + \left(\frac{\partial v}{\partial z}\right)^2}, \tag{2.58}$$

where N is the buoyancy frequency, u and v are the zonal and meridional components of the horizontal flow, respectively. The Richardson number expresses the importance of vertical variations in a horizontal shear flow. If $R_i > 0.25$, the flow is stable, that is, if the fluid is vertically displaced by a disturbance, it will either return to its initial position or perform periodic oscillations around the equilibrium position. For $R_i < 0.25$, the flow undergoes dynamical instability and if the $R_i < 0$ then the flow is convectively instable. Another interpretation of R_i is as the ratio of production of turbulence by buoyancy to the production of turbulence by wind shear (Nappo 2002). Typically, dynamical instability preceeds convective instability. Horizontal flow with large wind shear is more prone to instability than the flow with small wind shear.

2.15 Reynolds Number

The Reynolds number $\Re$ is an important dimensionless parameter that characterizes the nature of flow patterns in a fluid. It is defined as the ratio of inertial forces to viscous forces

$$\Re = \frac{\rho |\mathbf{u}| L}{\mu} = \frac{uL}{\nu}, \tag{2.59}$$

where L is the length scale of the flow, ρ is the mass density, and μ is the viscosity coefficient (or dynamic viscosity). $\Re$ can be used to identify whether the flow is laminar or turbulent. In a dynamic fluid $\Re$ is variable. Increasing $\Re$ means that intertial effects are becoming gradually more dominant over viscous effect and the flow thus has a tendency to transition to a turbulent state.

References

Chandrasekhar S (1960) Radiative transfer. Dover books on physics, Dover

Medvedev AS, Greatbatch RJ (2004) On advection and diffusion in the mesosphere and lower thermosphere: the role of rotational fluxes. J Geophys Res 109:D07104. doi:10.1029/2003JD003931

Nappo CJ (2002) An introduction to atmospheric gravity waves, International geophysics series, vol 85. Academic Press, Amsterdam

Yiğit E, Ridley AJ (2011) Role of variability in determining the vertical wind speeds and structure. J Geophys Res 116:A12305. doi:10.1029/2011JA016714

Chapter 3
Some Mathematical Basics: Conservation Laws

Equations of Atmospheric Physics

The conservation of energy would let us think that we have as much energy as we want. Nature never loses or gains energy. Yet the energy of the sea, for example, the thermal motion of all the atoms in the sea, is practically unavailable to us. In order to get that energy organized, herded, to make it available for use, we have to have a difference in temperature, or else we shall find that although the energy is there we cannot make use of it. There is a great difference between energy and available energy. The energy of the sea is a large amount, but it is not available to us
—R. Feynman (20th century A.D.)

Abstract The concepts of material derivative and reference frames are introduced. Conservation laws of physics are summarized and their application in the context of the governing equations of atmospheric dynamics is presented briefly. The set of nonlinear dynamical equations that describe the motion of an atmospheric parcel is called the Navier-Stokes equations. Various assumptions can be made to simplify the governing equations. Geostrophic wind approximation is presented, which is often used in lower and middle atmosphere dynamics. Finally, some vertical coordinate systems are discussed.

Keywords Material derivative · Conservation of energy · Conservation of momentum · Conservation of mass · Equations of atmospheric dynamics · Navier-Stokes equations · Vertical coordinate system

3.1 Introduction

We next will dwell on the fundamental equations of atmospheric dynamics. The basic question is what kind of laws or equations are needed in order to describe the changes that a parcel of air with volume δV undergoes. The atmosphere is described in the context of the continuum hypothesis (Sect. 2.3). An adequate description of the

© The Author(s) 2015
E. Yiğit, *Atmospheric and Space Sciences: Neutral Atmospheres*,
SpringerBriefs in Earth Sciences, DOI 10.1007/978-3-319-21581-5_3

Table 3.1 Topics of conservation laws and the associated conservation equations

Conservation law	Physics	Equation
Momentum conservation	Motion	Equation of motion
Energy conservation	Temperature	Thermodynamic energy equation
Mass conservation	Concentration	Continuity equation

changes a fluid parcel experiencing in its environment is described by three fundamental conservation laws of physics: momentum, energy, and mass. The momentum conservation is described by the infamous Newton's second law; the energy conservation is represented by the thermodynamic energy equation, following from the first law of thermodynamics; and the mass conservation is based on the equation of continuity. The underlying questions concerning an air parcel related to the conservation laws are:

- What processes controls the motion of air?
- How does the temperature change?
- How does mass concentration change?

These aspects are summarized in Table 3.1 in terms of the conservation laws.

3.2 Reference Frames in Fluid Dynamics

Conservation laws can be derived in general by considering the budgets of momentum, energy, and mass for an infinitesimal control volume δV. There are two main reference frames in fluid dynamics: *Eulerian* and *Lagragian*. The Eulerian frame of reference is fixed relative to the coordinate axis. Conservation will depend on the fluxes through the boundaries.

In the Lagrangian frame of reference, the control volume moves following the motion of the fluid, always containing the same fluid particles. This is a more useful method because the conservation laws can be stated in terms of a particular mass element. Also, this approach is necessary when one follows the time evolution of fields for various individual particles.

As we are interested in the total change, temporal and spatial changes have to be considered. All conservation laws can be used to describe these changes adequately. We first describe the total differential in the next section, then the three conservation laws will be briefly introduced and discussed in the context of the equations of atmospheric dynamics.

3.3 Material Derivative

In the atmosphere, any changes in a field variable can be associated with local changes and changes due to a motion. The material derivative is the appropriate operator that quantifies the total change in a field variable. In general, the total derivative is defined as

$$\frac{d}{dt} \equiv \underbrace{\frac{\partial}{\partial t}}_{local} + \underbrace{\mathbf{v} \cdot \nabla}_{advective} , \tag{3.1}$$

and is composed of a local $\frac{\partial}{\partial t}$ and an advective term $\mathbf{v} \cdot \nabla$. Sometimes, this operator is called the *substantial* or the *total* derivative or even the convective derivative. The three-dimensional velocity $\mathbf{v}$

$$\mathbf{v} = u\,\hat{\mathbf{i}} + v\,\hat{\mathbf{j}} + w\,\hat{\mathbf{k}},$$
$$\mathbf{v} = (u, v, w), \tag{3.2}$$

is composed of the zonal (u), meridional (v), and vertical components (w), where $\hat{\mathbf{i}}, \hat{\mathbf{j}}$, and $\hat{\mathbf{k}}$ are the Cartesian unit vectors. The total derivative is then given in components

$$\frac{d}{dt} = \frac{\partial}{\partial t} + \underbrace{u\frac{\partial}{\partial x} + v\frac{\partial}{\partial y} + w\frac{\partial}{\partial z}}_{advective}, \tag{3.3}$$

which shows the zonal, meridional, and the vertical advection terms. In planetary atmospheres, the total change in a field parameter depends on local changes at a fixed point plus changes following the motion. For a stationary process, the total change is given by Euler's description of a fluid. This means that the total change in a fluid property is given by the local rate of change at a fixed point:

$$\frac{d}{dt} = \frac{\partial}{\partial t}. \tag{3.4}$$

In general, the total differential can be understood as a superposition of the rate of change of a field variable following the motion and its rate of change at a fixed point (local). For example, the total differential of temperature T is obtained by applying the operator (3.1) on T.

$$\frac{dT}{dt} = \frac{\partial T}{\partial t} + u\frac{\partial T}{\partial x} + v\frac{\partial T}{\partial y} + w\frac{\partial T}{\partial z}. \tag{3.5}$$

Then, the total temperature change results from local changes plus the advection of temperature in three-dimensional space. To illustrate the phenomenon of total derivative, consider the temperature of water in a closed volume element in a lake.

Temperature rise can occur locally because of the absorption of solar radiation, i.e., $\frac{\partial T}{\partial t} > 0$. If there is a stream in the lake, the motion of water from other parts of the lake will cause an advective temperature change, i.e., $|(\mathbf{v} \cdot \nabla)T| > 0$. Advective changes become most significant for large speeds in the presence of spacial inhomogeneity in the fluid properties. Consider the example of the volume element in a lake at constant temperature. In this case, even if there is a strong stream within the lake, advection will not produce a temperature change if $\nabla T = 0$. In a realistic atmosphere, a significant degree of spatial temperature gradients exists in the presence of large winds.

In cases when the spherical geometry has to be assumed, the total derivative can be expressed in terms of spherical polar coordinates (2.2). For instance, the temperature $T = T(r, \phi, \theta)$ varies with radial distance r, longitude ϕ, and latitude θ. We then have for the total derivative of the temperature

$$\nabla T = \frac{\partial T}{\partial r}\,\hat{\mathbf{r}} + \frac{1}{r}\frac{\partial T}{\partial \theta}\,\hat{\boldsymbol{\theta}} + \frac{1}{r\cos\theta}\frac{\partial T}{\partial \phi}\,\hat{\boldsymbol{\phi}} \tag{3.6}$$

3.4 Rotational Effects

To apply in a planetary atmosphere context, the total differential has to be transformed into a rotating frame of reference. This refers the motion to a reference frame that is rotating with the planet. On the large-scale, motions of the atmosphere and oceans are all affected by rotation and these effects give rise to new phenomena. For this transformation, a link between the total differential of a vector $\mathbf{A}$ in an inertial frame and in a rotating system must be formulated. It is given by

$$\left(\frac{d\mathbf{A}}{dt}\right)_r = \frac{d\mathbf{A}}{dt} + \boldsymbol{\Omega} \times \mathbf{A}, \tag{3.7}$$

where subscript r denotes the rotating system and $\boldsymbol{\Omega} = (\Omega_x, \Omega_y, \Omega_z)$ is the angular velocity vector. Then $|\boldsymbol{\Omega}| = \Omega = 2\pi/\tau$, which is variable among the different planets. For Earth, the rotational rate is $\Omega = 7.292 \times 10^{-5}$ rad s^{-1} and Mars has a slightly slower rate of $\Omega = 7.077 \times 10^{-5}$ rad s^{-1}. In comparison Jupiter, the largest Solar System planet, has a rotation rate of $\Omega = 1.774 \times 10^{-4}$ rad s^{-1}. The components of the angular velocity are given by

$$\boldsymbol{\Omega} = \begin{pmatrix} \Omega_x \\ \Omega_y \\ \Omega_z \end{pmatrix} = \begin{pmatrix} 0 \\ \Omega\cos\theta \\ \Omega\sin\theta \end{pmatrix}. \tag{3.8}$$

The importance of rotational effects can be judged from the *Rossby number* R_o, which expresses the ratio of the characteristic horizontal wind speed U and the Coriolis terms

$$R_o \equiv \frac{U}{fL}, \tag{3.9}$$

where $f = 2\Omega \sin\theta$ is the Coriolis parameter and L is a characteristic length scale.

3.5 Conservation of Momentum

Momentum conservation is given by Newton's second law of motion

$$\frac{d\mathbf{v}}{dt} = \sum_i \mathbf{F}_i, \tag{3.10}$$

where the total acceleration is given by the sum of all the forces per unit mass $\mathbf{F}_i$. If the motion is viewed in a rotating system, then apparent forces arise. Therefore, we have to transform Newton's law to a rotating coordinate system first. Let us start by applying the operation in (3.7) to the position vector first

$$\left(\frac{d\mathbf{r}}{dt}\right)_r = \frac{d\mathbf{r}}{dt} + \mathbf{\Omega} \times \mathbf{r}, \tag{3.11}$$

which can be written in terms of the velocity

$$(\mathbf{v})_r = \mathbf{v} + \mathbf{\Omega} \times \mathbf{r}, \tag{3.12}$$

stating that the absolute velocity of an object on rotating Earth is given by the velocity of the object relative to Earth plus the velocity due to the rotation of Earth. Combining (3.10) with the velocity in a rotating frame (3.12) yields

$$\left(\frac{d\mathbf{v}}{dt}\right)_r = \frac{d\mathbf{v}}{dt} + 2\mathbf{\Omega} \times \mathbf{v} - \Omega^2 \mathbf{r}. \tag{3.13}$$

Equation (3.13) states that the acceleration of a fluid parcel following the motion is given by the acceleration following the relative motion in the rotating frame plus the Coriolis resulting from the relative motion in the rotating frame plus the centripetal acceleration caused by coordinate rotation. If we assume that the real forces in a rotating frame are given by the pressure gradient force, gravitation (combined with centripetal term), and some friction force $\mathbf{F}_f$, then the equation of motion becomes

$$\frac{d\mathbf{v}}{dt} = -\frac{1}{\rho}\nabla p - 2\mathbf{\Omega} \times \mathbf{v} + \mathbf{g} + \mathbf{F}_f. \tag{3.14}$$

In dynamic meteorology, Eq. (3.14) is the most general form of momentum equation. In an ionized atmosphere, this equation has to incorporate additional terms resulting from the interaction of neutrals with ions.

3.6 Conservation of Energy

The total energy E found within a fluid volume is composed of the internal energy U and the kinetic energy E_k

$$E_t = E_k + U. \tag{3.15}$$

The associated total energy density is

$$\varepsilon_t = \varepsilon_k + \varepsilon. \tag{3.16}$$

Energy conservation concerns the rate of change of energy in a volume element due to internal and external processes and is represented by the first law of thermodynamics (2.17). That is, the change in the internal energy of a system is given by the sum of the changes induced by heat flow and work done on the system. Taking the energy conservation in terms of the specific quantities as in Eq. (2.30)

$$dq_s = d\varepsilon + pd\nu, \tag{3.17}$$

which can be written, using the condition that the energy of an ideal gas depends on the the the temperature, $d\varepsilon = c_v dT$, as

$$dq_s = c_\nu \, dT + p \, d\nu. \tag{3.18}$$

Expanding the exact differential $d\nu$ in terms of its partial differential, using the ideal gas law and the property

$$c_p - c_\nu = R, \tag{3.19}$$

lead to

$$dq_s = c_p \, dT - \frac{RT}{p} \, dp. \tag{3.20}$$

The formulation of temperature tendency gives one form of the energy equation

$$\frac{dT}{dt} = \frac{1}{c_p} \left(\frac{dq_s}{dt} + \frac{RT}{p} \frac{dp}{dt} \right). \tag{3.21}$$

3.7 Conservation of Mass

Conservation of mass simply states that the rate of change of mass in a volume element is the difference between the mass flux into the volume element and mass flux out of the volume. This conservation law is expressed by the continuity equation

Fig. 3.1 Illustration of
divergence (*left*) and
convergence (*right*) with
respect to a *circular* region
shown in *black*

$$\frac{\partial \rho}{\partial t} + \mathbf{\nabla} \cdot (\rho \mathbf{v}) = 0, \tag{3.22}$$

where $\mathbf{\nabla} \cdot (\rho \mathbf{v})$ expresses mass outflow per unit volume and accordingly, $-\mathbf{\nabla} \cdot (\rho \mathbf{v})$ is the mass inflow per unit volume. Equation (3.22) is the *mass divergence* form of the continuity equation. It states that the local density change $\frac{\partial \rho}{\partial t}$ is given by the mass inflow/outflow, that is, by mass convergence/divergence, which is illustrated in Fig. 3.1. A local decrease in density $\frac{\partial \rho}{\partial t} < 0$ would imply mass divergence.

Using the total differential operator, an alternative form of the continuity equation, the *velocity divergence* form is obtained

$$\frac{1}{\rho} \frac{d\rho}{dt} + \mathbf{\nabla} \cdot \mathbf{v} = 0, \tag{3.23}$$

stating that the fractional rate of increase of density is equal to velocity convergence (that is minus divergence).

3.8 Incompressible Flow

In general, in geophysical flows, density is a function of pressure and temperature, i.e., $\rho = \rho(p, T)$. Often, the density variations throughout the volume of the fluid and throughout its motion can be invariant. In this case, any compression or expansion is negligible and the flow is to said to be *incompressible*. In other words, we neglect the change of density by the change of pressure. This assumption helps simplify the dynamical equations.

3.9 Governing Equations of Atmospheric Dynamics

The equations of atmospheric dynamics governing the motion of an air parcel are based on the conservation of energy, momentum, and mass discussed in Sects. 3.5–3.7. Combined with the equation of state for an ideal gas (2.12), an adequate closed set of Eqs. (3.14), (3.21), and (3.22) are obtained from the conservation laws that describe the energy, momentum and mass transfer in planetary atmospheres.

In general, the fundamental equations of fluid dynamics that describe atmospheric flows in detail are called *Navier-Stokes Equations*.[1] The basic form of the Navier-Stokes equations include the pressure gradient force and viscous forces (Anderson 1995).

$$\frac{d\mathbf{u}}{dt} = -\frac{1}{\rho}\nabla p + \frac{\mu}{\rho}\nabla^2\mathbf{u}, \tag{3.24}$$

where $\mu = \nu\rho$ is the coefficient of viscosity expressed in terms of the kinematic viscosity ν. In Eq. (3.24), it is assumed that the spatial temperature differences are relatively small so that the dynamic viscosity μ is uniform over the fluid. However, if significant temperature differences are present in a fluid, the viscosity coefficient can vary with space. Therefore, $\nu\nabla^2\mathbf{u}$ should be replaced by $\nabla[(\mu/\rho)\nabla \cdot \mathbf{u}]$ (Batchelor 2000). The Navier-Stokes equation of motion can be expanded by including more body forces, such as, the gravitational forces, Coriolis force, ion-neutral coupling terms. Then, a more comprehensive form of the momentum equation is

$$\frac{d\mathbf{u}}{dt} = -\frac{1}{\rho}\nabla p + \nu\,\triangle\mathbf{u} - 2\mathbf{\Omega} \times \mathbf{u} + \mathbf{g} - \nu_{ni}(\mathbf{u} - \mathbf{v}_i), \tag{3.25}$$

where $\triangle \equiv \nabla^2$ and the last expression on the right hand-side represents the ion drag with the neutral-ion collision frequency ν_{ni} and the plasma (ion) velocity $\mathbf{v}_i$.

The density of a chemical species in the atmosphere is described by the continuity equation. However, change of concentration can result from chemical reactions (local effects) as well as from transport processes. Applying the continuity to the number density of a species $n_i = \rho_i/m_i$ gives

$$\frac{\partial n_i}{\partial t} + \nabla \cdot (n_i\mathbf{u}) = S_i, \tag{3.26}$$

where S_i expresses a source term, which incorporates production P_i and loss processes L_i as $S_i = P_i - L_i$ and $\nabla (n_i\mathbf{u})$ is the transport of the species i by the atmospheric circulation $\mathbf{u}$. The unit of the source term is then concentration per unit time.

For small vertical air parcel displacements, Boussinesq approximation (see Sect. 5.6.1) can be assumed, which is appropriate for incompressible flows. Under these circumstances, the fluid has a constant density following the motion

$$\frac{d\rho}{dt} = 0 \tag{3.27}$$

Then, the continuity equation for incompressible flow is given by

$$\nabla \cdot \mathbf{u} = 0. \tag{3.28}$$

---

[1]Named after two scientists, M. Navier and G. Stokes.

The energy equation for the incompressible flow is given by

$$\frac{\partial T}{\partial t} = -\frac{1}{\rho c_p}(\mathbf{u} \cdot \nabla)T + \alpha \nabla^2 T + \frac{\nu}{2c_p}\Phi_D + J_R, \tag{3.29}$$

where first term is the advection of temperature, second is heat conduction, third is the dissipation due to fluid stress with the dissipation function Φ_D, and J_R is radiative heating term. Finally, the equation of state is given by

$$p = \rho R_* T. \tag{3.30}$$

Equations (3.25), (3.28)–(3.30) represent the incompressible equations of atmospheric dynamics. Depending on the nature of problem, additional source or sink terms have to be considered in these equations. These equations are overall three-dimensional nonlinear time-dependent partial differential equations that can either be solved using numerical methods or using appropriate approximations. For example, when the pressure gradient force approximately balances the Coriolis force and all other forces are negligible, the atmosphere is said to be in geostrophic balance. Then Eq. (3.25) becomes

$$\frac{1}{\rho}\nabla p \approx -2\boldsymbol{\Omega} \times \mathbf{u}. \tag{3.31}$$

The horizontal momentum equations then are

$$\frac{1}{\rho}\frac{\partial p}{\partial x} \approx 2v\Omega \sin\theta - 2w\Omega \cos\theta \tag{3.32}$$

$$\frac{1}{\rho}\frac{\partial p}{\partial y} \approx -2u\Omega \sin\theta \tag{3.33}$$

Provided that the horizontal flow speeds are much larger than the vertical flows, i.e., $u \gg w$ and $v \gg w$, the flow can be assumed to be predominantly horizontal, then the zonal and the meridional geostrophic winds are given by

$$v_g = \frac{1}{f\rho}\frac{\partial p}{\partial x} \tag{3.34}$$

$$u_g = -\frac{1}{f\rho}\frac{\partial p}{\partial y} \tag{3.35}$$

where $f = 2\Omega \sin\theta$ is the Coriolis parameter. At a typical midlatitude of 45°N $f \approx 0.001\,\text{rad s}^{-1}$.

3.10 Vertical Coordinate System

The atmosphere-ionosphere system exhibits significant variations with altitude. The choice of an appropriate vertical coordinate system is very important in terms of the representation of the governing equations of atmospheric dynamics. Some common coordinate systems are altitude (z), pressure or isobaric (p), sigma (σ), and isentropic vertical coordinate systems. The σ-coordinate system uses the ratio of the pressure to the surface pressure, $p(z)/p_s$. The isentropic system is based on surfaces of constant potential temperature θ (Sect. 2.11). Furthermore, there are hybrid systems, such as, sigma-pressure and sigma-altitude vertical coordinates. The different coordinate systems can provide different advantages and simplifications in the treatment of the fluid dynamical equations.

The most intuitive vertical coordinate system is probably the altitude coordinate. However, the governing equations of atmospheric dynamics are often represented by coordinates other than the altitude coordinate. The choice of a vertical coordinate system is a central aspect of atmospheric global modeling. There are a number of aspects to be considered in choosing an appropriate vertical coordinate. For example, the treatments of nonhydrostatic processes and terrains (orography) are two special cases. The altitude and pressure coordinates both intersect the surface topography while the sigma-pressure or sigma-altitude are terrain following coordinates, that is, the model vertical layers do not intersect the orography. On the other hand, both the pressure and sigma-pressure coordinates assume hydrostatic balance (2.31). In order to represent nonhydrostatic fluid dynamical equations, an altitude coordinate system ought to be used. These properties of the different vertical coordinates are summarized in Table 3.2.

The hydrostatic equilibrium is an adequate assumption in modeling only the large-scale atmospheric processes in the absence of strong perturbations. However, hydrostatic assumption can easily break down in relatively smaller atmospheric scales (e.g., mesoscale and smaller) and under disturbed atmospheric conditions.

Another useful coordinate system is the so-called log-pressure altitude system, which is commonly used by meteorologists in atmospheric models in order simplify

Table 3.2 Various vertical coordinate systems used for the representation of the equations of atmospheric dynamics

Coordinate	Nonhydrostatic	Terrain-following
Altitude	✓	✗
Pressure	✗	✗
Isentropic	✗	✗
Sigma	✗	✗
Sigma-pressure	✗	✓
Sigma-altitude	✓	✓

Here they are characterized according the appropriateness for nonhydrostatic and terrain-following processes

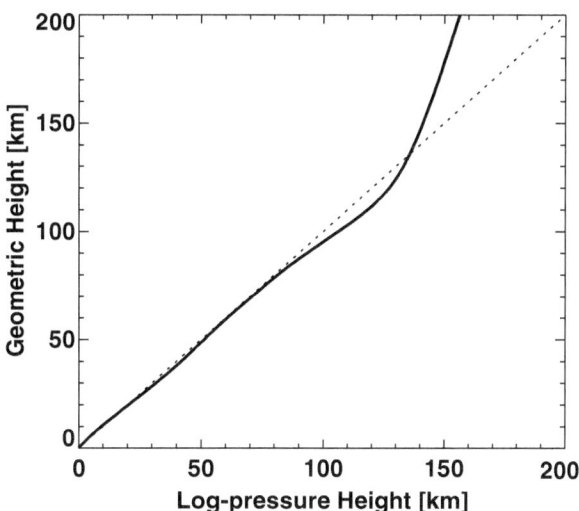

Fig. 3.2 Geometrical height (*dotted*) and log-pressure height (*thick-solid*) line based on NRL-MSISE data for a Northern Hemisphere midlatitude in January

model equations. The associated log-pressure altitude is given by

$$z_p = -H_m \ln\left(\frac{p}{p_s}\right), \tag{3.36}$$

where $p_s \approx 1000$ mb is the pressure at the sea level, pressure $p = p_s \exp(-z/H)$, and H is a mean scale height given by $R_* T_s / g_0$ that is typically taken to be constant ($H = 7$ km) in middle atmosphere studies. The z_p scales nearly linearly with the geometric height. Figure 3.2 illustrates the relation between the log-pressure height z_p and the geometrical height z. In the lower and middle atmosphere, the difference between them is negligible but in the upper atmosphere, z_p can significantly depart from the actual geometrical height due primarily to stronger temperature variations.

References

Anderson JD (1995) Computational fluid dynamics. McGraw-Hill, New York

Batchelor GK (2000) Introduction to fluid dynamics. Cambridge University Press, Cambridge Mathematical Library Series, Cambridge

Chapter 4
Earth's Atmosphere and Geospace Environment

Man must rise above the Earth to the top of the atmosphere and beyond for only thus will he fully understand the world in which he lives.

—Socrates (5th century B.C.)

Abstract The Sun-Earth system displays a typical relationship of a terrestrial planet around its parent star. This connection between Sun and Earth is more clearly represented by the space weather systems: Sun, heliosphere, magnetosphere, and atmosphere-ionosphere. Without the investigation of these systems and their interactions with each other, the effects of Sun on Earth's atmosphere cannot be fully understood. Overall, Earth is influenced by Sun directly and indirectly via radiative, thermal, dynamical, and electrodynamical processes. Earth's atmosphere is stably stratified and consists of different regions of varying chemical and physical properties. In particular, Earth's upper atmosphere-ionosphere system is controlled by solar and magnetospheric processes from above and by internal waves from below. Various space weather systems from Sun to the lower atmosphere of Earth, the troposphere, are briefly discussed.

Keywords Troposphere · Stratosphere · Mesosphere · Thermosphere · Ionosphere · Atmosphere-ionosphere · Magnetosphere · Heliosphere · Sun

4.1 Sun-Earth Connection

In this chapter we will present an overview of the Sun-Earth system starting from Sun down to Earth's lower-most atmosphere, the troposphere. Sun's and Earth's atmosphere and the regions in between are sometimes referred to as space weather systems. This is because the radiative and thermal effects of Sun propagate through large distances down to Earth's surface. Our approach is also thought to emphasize that between the space weather systems there is a large degree of coupling.

© The Author(s) 2015
E. Yiğit, *Atmospheric and Space Sciences: Neutral Atmospheres*,
SpringerBriefs in Earth Sciences, DOI 10.1007/978-3-319-21581-5_4

4.2 Sun

Our Sun is formed about five billion years ago and is the main supplier of life on our planet. It consists mainly of hydrogen ($\sim$73 %) and Helium ($\sim$25 %). It also has other elements, such as, carbon, oxygen, iron, and magnesium in relatively smaller amounts. It harnesses its energy from fusion reaction, which transforms hydrogen to the heavier deuterium and subsequently to Helium. This processes is extremely exothermic and thus releases energy to the environment. With a radius of $\sim$7 $\times$ 10^8 m, our Sun is about 100 times larger than Earth and is just like an enormous nuclear power plant sitting (nearly) in the center of the Solar System. It emits continuous stream of radiation that interacts with the neutral atmosphere of planets in the Solar System. The associated radiation consists of a broad range of electromagnetic waves, ranging from radio waves to very high-energy γ-rays. The solar radiation especially in the UV- and EUV-range plays the primary role of ionization in Earth's atmosphere. Table 4.1 lists four different ranges of the electromagnetic spectrum from the visible to X-rays as a function of wavelength. In general, ionization in planetary atmospheres depends on the particular distribution of chemical species.

Owing to a magnetohydrodynamic dynamo in its core, Sun has a strong intrinsic magnetic field, which is responsible for various dynamic features on its surface. Sunspots are cooler and thus darker areas on the surface of the Sun that owe their existence to strong magnetic fields. In the 19th century, H. Schwabe discovered that the number of sunspots vary with time and it reaches a maximum every 11 years, which is called the 11-year solar cycle. The sunspots typically occur in groups. Within groups each pair of sunspots has opposite polarity. Every solar cycle this polarity reverses. Figure 4.1 presents the solar cycle variations from 1963 to-date based on the $F_{10.7}$ index, which is a good proxi for solar activity. The black and red lines are the hourly averaged and a 30-day running mean values of the $F_{10.7}$ index, respectively. The 11-year variations are clearly seen but there are also short-term variations in the solar activity. During solar maximum periods, $F_{10.7}$ can exceed 250×10^{-22} W m^2 Hz^{-1} and during solar minimum phase it is as low as 70×10^{-22} W m^2 Hz^{-1}.

The solar atmosphere consists of the photosphere, chromosphere and the corona. Photosphere is the lower-most layer of the solar atmosphere and is about 6000 K. The chromosphere is only slightly warmer than the photosphere but the temperature gradually increases till the corona, where the temperatures are in the order of a million K. Thus, the transition region between the solar chromosphere and corona has a very strong temperature gradient as the temperature increases two orders of magnitude

Table 4.1 Portion of the solar electromagnetic spectrum from the visible to X-rays

Electromagnetic radiation	Wavelength λ (nm)
Visible	380–750
UV	100–330
EUV	10–100
X-rays	<10

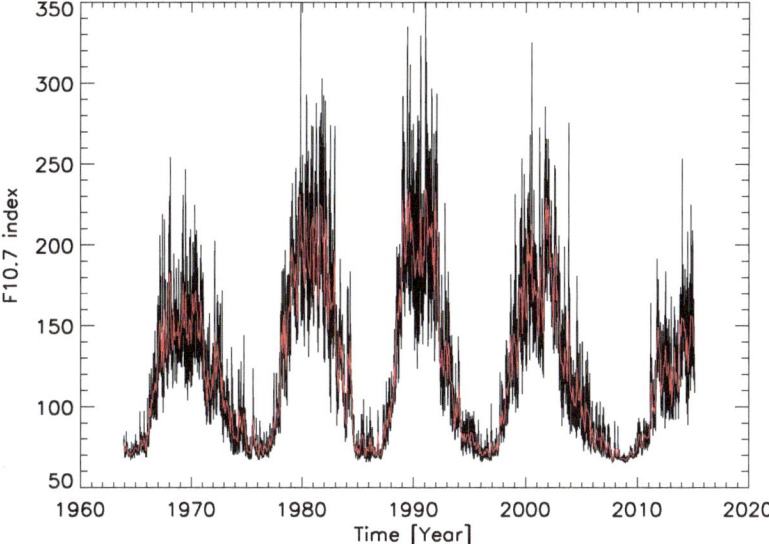

Fig. 4.1 Solar F10.7 cm flux variations from 1963–2015. Data obtained from NASA's OMNI data base. The *black* and *red lines* are the hourly averaged and a 30-day running mean values of the $F10.7$ index

($\sim 10^4$ to $\sim 10^6$ K) over a distance of few hundred km. The corona is highly ionized and extends far into the interstellar space. The Extreme ultraviolet imaging telescope on board the Solar and Heliospheric Observatory (SOHO) continuously images Sun's atmosphere at various frequencies as shown in Fig. 4.2. Here images taken at (a) 17.1 nm, (b) 19.5 nm, (c) 28.4 nm, and (d) 30.4 nm are presented, highlighting features with temperature at around 1, 1.5, 2 million K, and 60,000–80,000 K, respectively. Increasing temperatures indicate that we are looking higher in the atmosphere of Sun. SOHO provides an unprecedented monitoring of the solar activity at real time.

4.3 Heliosphere

The heliosphere is the extension of Sun's environment into the Solar System and beyond. In general, we can call it a "starsphere" (Kallenrode 2004) which is void in space and is characterized by the solar (star) wind and magnetic flux. The solar wind carries the frozen-in solar magnetic field, which forms the interplanetary magnetic field (IMF) $\mathbf{B}_{IMF}$. The heliosphere can extend as much as 100 AU.

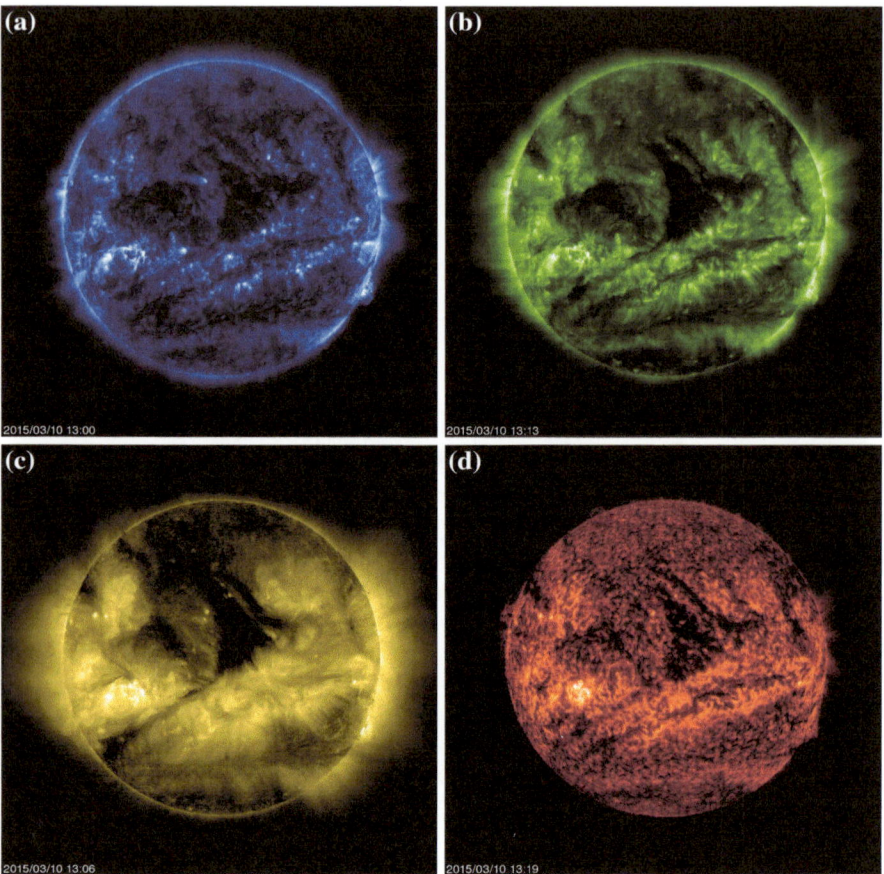

Fig. 4.2 Images of the solar atmosphere at several wavelengths taken with the. Extreme ultraviolet imaging telescope aboard the Solar and Heliospheric Observatory (SOHO) at **a** 17.1 nm (1 million K), **b** 19.5 nm (1.5 million K), **c** 28.4 nm (2 million K), **d** 30.4 nm (60,000–80,000 K). Courtesy of NASA and ESA

The solar wind is a collection of high-energy supersonic charged particles, consisting of ions and electrons. It originates at Sun because of the pressure difference between the solar atmosphere (corona) and the surrounding interstellar space. The outward directed enormous pressure gradient then drives the plasma away from Sun into the solar system. The basic understanding of the solar wind formation can be obtained from theoretical models that treat the solar coronal plasma in the gravitational field of Sun. Such a model is based on conservation equations in a spherically symmetric geometry with its origin at the center of Sun.

Away from Sun the IMF lines have a spiral structure frozen in a radially expanding solar wind. The effects of the solar wind can be detected well past the orbit of Earth. The solar wind in the vicinity of Earth (i.e., at 1 AU) is hot, fast, and quasi-neutral.

Typical electron and proton temperatures are in the order of 10^5 K, which is about one hundred times hotter than Earth's hottest atmospheric region, the thermosphere. Ionized hydrogen forms the main constituent of the solar wind. The quiet-time solar wind speed is about 400–500 km s^{-1} with a magnetic field of the order of 10^{-9} T (1 nT).

4.4 Magnetosphere

Owing to an active dynamo in its core, Earth possesses a strong intrinsic magnetic field. This field is a strong dipole, meaning that the dipole term is dominant in the presence of higher-order terms, and is tilted and offset with respect to the rotational axis of Earth. In Earth's environment the resultant magnetic field is a superposition of the intrinsic field and a field resulting from external sources, such as the IMF. Therefore, the resultant magnetic scalar potential V_m is a superposition of the intrinsic term and an external term.

The magnetosphere is the magnetic field region surrounding Earth, where the properties of the plasma are controlled by the geomagnetic field. The terrestrial magnetosphere can be divided into two components: the inner and outer magnetosphere. The dipole region of the magnetosphere close to Earth is called the inner magnetosphere. The hydromagnetic waves, such as Alfvén waves, play an important role in exchanging information between the magnetosphere and ionosphere. The non-dipolar region of the magnetosphere at the nightside (anti-sunward direction) starting at about 6.5 Earth radii is defined as the outer magnetosphere. Overall, the magnetosphere can extend up to 10 Earth radii in the sunward direction and up to few hundred r_e in the anti-sunward direction, forming a long tail (magnetotail).

Earth's magnetosphere is influenced by the impact of the continuous stream of solar wind originating at Sun. Interaction of the solar wind with the terrestrial magnetic field leads to a compression in the sunward direction. This interaction results in various current systems within the magnetosphere, such as, the ring current, tail current, and field-aligned currents, which are ultimately mapped down to the ionosphere, contributing to ionospheric currents.

Figure 4.3 shows a sketch of the magnetosphere. On the left hand-side the arrival of the solar wind is shown with speed v_x, which carries the solar magnetic field, which is the interplanetary magnetic field (IMF), $\mathbf{B}_{IMF}$. The solar magnetic field is said to be frozen-into the solar wind. Here, an idealized case of southward IMF is shown, which is able to connect to the northward directed geomagnetic field. This connection between the solar wind magnetic field and Earth's magnetic field is called *reconnection*. The *magnetopause* separates Earth's magnetic field from the magnetized solar wind. A comprehensive review of geomagnetism and magnetospheric physics can be found in the work by Stern (2002).

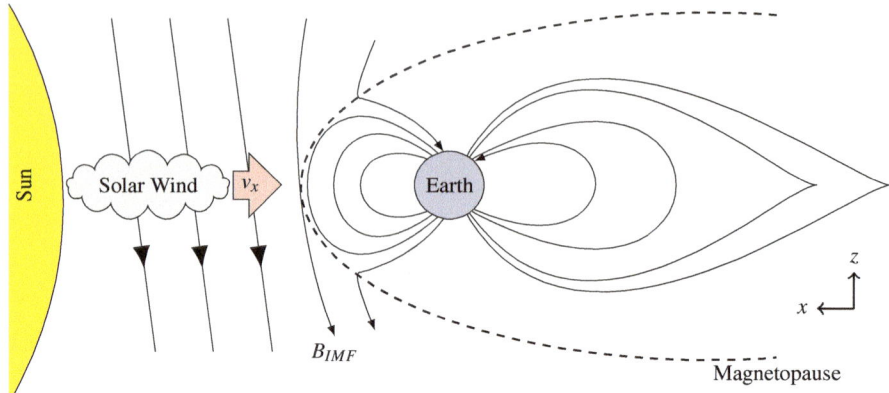

Fig. 4.3 A simple illustration of Earth's magnetosphere environment

4.5 Plasmasphere

The plasmasphere is the region of cold, dense ($\sim 10^3$–10^4 cm^{-3}) plasma immediately surrounding and co-rotating Earth. It mostly consists of helium and hydrogen particles, but also oxygen, which has managed to escape Earth's ionosphere. Outer boundary of the plasmasphere is the plasmapause. The location of the Van Allen[1] radiation belts and the ring current approximately coincides with the plasmasphere. The radiation belts have high-energy radiation that can damage space instrumentation (e.g., satellites) and harm astronauts.

4.6 The Atmosphere-Ionosphere System

The atmosphere, a thin layer of gas surrounding the solid Earth, is composed of various layers or regions, forming the neutral atmosphere. At higher altitudes, the neutral atmosphere is increasingly ionized primarily by the incoming solar EUV/UV radiation, producing a partially ionized plasma environment, the ionosphere, consisting of ions and electrons that are characterized by the associated number densities n_i and n_e. Thus, the atmosphere and ionosphere are indeed not two separate regions. The ionosphere is simply the ionized portion of the upper atmosphere, meaning that it coexists with the upper atmosphere.

The term "atmosphere-ionosphere" system has been increasingly used by the scientific community to describe coupling processes in Earth's whole atmosphere system (e.g., Abdu et al. 2006; Pancheva and Mukhtarov 2011; Jin et al. 2011; Pancheva et al. 2012; Laštovička 2013; Liu et al. 2014; Yiğit and Medvedev 2015).

---

[1] Named after the prominent physicist James Alfred Van Allen who discovered this region of trapped plasma.

One implication of such an expression is that these two regions are inseparable in terms of their scientific investigations. The international symposium on "Coupling Processes in the Atmosphere-Ionosphere System" is jointly organized by IAGA, ICMA, SCOSTEP at the IUGG meeting in Prague in summer 2015 in order to bring new insights into the understanding of the coupling processes in the atmosphere-ionosphere system.

The vertical structure of Earth's neutral atmosphere is described primarily by the variations of the neutral temperature with height. While the electron density distribution defines the vertical structure of the ionosphere. Next, I will briefly describe the different regions in the atmosphere-ionosphere system.

4.7 Neutral Atmosphere

The neutral atmosphere extends from ground to $\sim$500–900 km, depending on the solar and geomagnetic activities. It is divided into the troposphere, stratosphere, mesosphere, and thermosphere as seen in Fig. 4.4, which are separated from each other by layers including the suffix "pause", i.e., the tropopause, stratospause, and mesopause layers. These layers coincide with the temperature minima/maxima in the atmosphere and indicate a change in the vertical temperature gradient.

4.7.1 Troposphere

The troposphere, extending from Earth's surface to $\sim$10–15 km, is the lower boundary of the atmosphere, that is, the interface between Earth's surface and the middle atmosphere. It is the region where the lower atmospheric weather takes place. There is a significant amount of mixing, convective activity and latent heat effects in this region. It contains about 80 % of the total atmospheric mass, nearly all water vapor and clouds. The temperature drops with height in the troposphere (i.e., positive lapse rate, $\Gamma > 0$) until a temperature minimum is reached at the tropopause. The strong temperature decrease provides instability processes for wave generation in the lower atmosphere.

4.7.2 Stratosphere

The stratosphere, which is separated from the troposphere by the tropopause, extends from about 10–15 km up to 50 km and the temperature increases rapidly as a result of radiative balance, e.g., due to the absorption of solar EUV by the ozone molecules. Temperature increases, i.e., $\Gamma < 0$, which favors static stability, continues until the stratopause, a local temperature maximum. In the stratosphere, internal waves

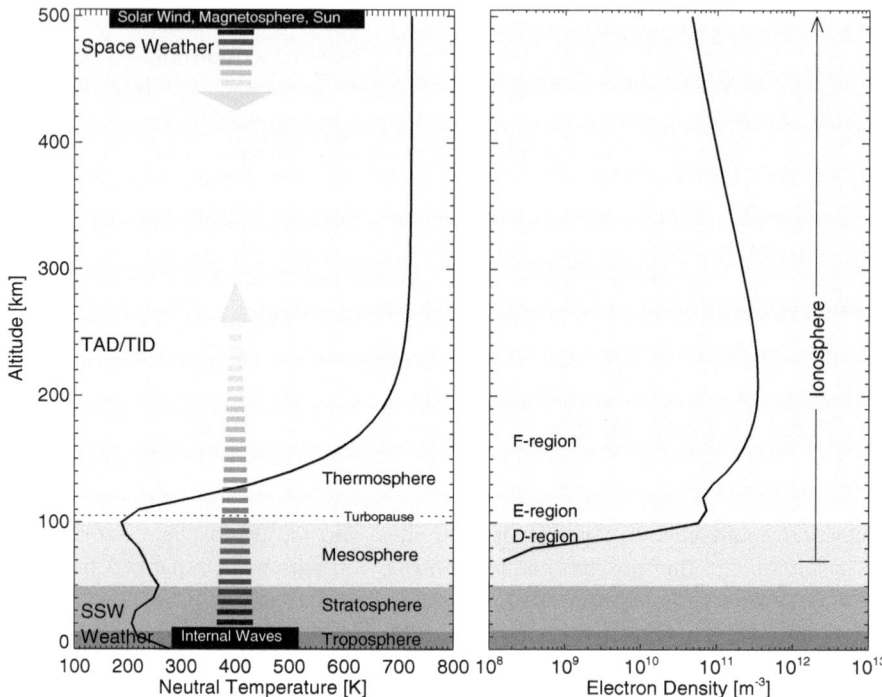

Fig. 4.4 Earth's atmosphere-ionosphere system represented with the neutral atmosphere in the *left* and ionosphere on the *right panels*. Data for the neutral temperature (*left*) and plasma (electron) density (*right*) mean profiles are taken from MSISE-90 and IRI2012 models, respectively. The turbopause is marked at ∼105 km. The neutral atmosphere regions and ionospheric regions are shown as well. Physical processes are written depending on their approximate occurrence altitudes: weather, SSW (sudden stratospheric warming), TAD/TAD (traveling atmospheric disturbances/traveling ionospheric disturbances), and space weather. Processes related to internal waves and solar wind, magnetosphere, Sun are denoted within *black boxes* to describe the influence from below and above, respectively, on the atmosphere-ionosphere system. Adopted from (Yiğit and Medvedev 2015, Fig. 1)

grow in amplitude and generally larger scale waves, such as inertio-gravity waves and planetary waves interact with the mean flow. This is also the region, where the zonal mean circulation exhibits a quasi-biennial oscillation (QBO). During Northern Hemisphere winter periods, provided that sufficient planetary wave activity is present, sudden stratospheric warmings occur in the polar stratosphere.

4.7.3 Mesosphere

The mesosphere extends from the stratopause (∼50 km) up to the 90–100 km, where the mesopause is located. The altitude of the mesopause is largely variable because the mesosphere is influenced dynamically by internal waves, primarily by solar tides and small-scale gravity waves. The summer mesopause is also the coldest place in Earth's atmosphere, where temperatures down to 130 K can be found. According to radiative arguments, the summer mesopause is expected to be warmer than the winter mesopause. However, the meridional temperature gradient is reversed by the momentum deposition primarily by gravity waves. This phenomenon is sometimes called the *mesopause anomaly*. The stratosphere and the mesosphere together make up the *middle atmosphere*.

4.7.4 Thermosphere

The thermosphere forms the upper most region of the neutral atmosphere and extends from the mesopause up to 500 km, depending largely on the solar activity and geomagnetic disturbances. It is also the hottest region in Earth's atmosphere with temperatures ranging from 600 to >1000 K, caused by the absorption of the solar radiation. The dominant cooling mechanism is typically the downward thermal conduction. In general, in planetary atmospheres, the thermosphere marks the hottest place.

Because the thermosphere is greatly influenced from below (internal waves) and from above (e.g., solar wind, particle precipitation, Sun) at different temporal and spatial scales, it possesses a significant degree of variability. The 11-year solar cycle has a profound effect on the energy budget of the thermosphere. As large number of satellites have their orbit within the thermosphere, characterization of understanding thermospheric variability is crucial for better planning of observational missions and for the safety of space missions. Thermospheric density is much smaller than that is found in the lower atmosphere. In this region, diffusive separation gradually dominates over the turbulent mixing that takes place at lower altitudes and therefore individual species have their own scale heights (Eq. 2.37). At greater heights, the atomic oxygen is the major species. This means that the different chemical species exhibit different altitude profiles.

The thermosphere coincides with the ionosphere (Sect. 4.8), which is the ionized portion of the neutral atmosphere. The thermosphere-ionosphere is a region of natural plasma laboratory, in which ion-neutral coupling processes can be investigated. The coupling between the plasma and neutrals is greatly modulated by the magnetosphere, the natural plasma shield around Earth.

4.7.5 Exosphere

The thermosphere is separated by the exobase ($\sim$500 km on Earth and $\sim$300 km on Mars) from the exosphere, where particles with significant amount of kinetic energy can escape from Earth's gravitational field. Above the exobase the mean free path exceeds the atmospheric scale height. The Moon and Mercury have their exobases at their surface. The exosphere is an important region for loss of atmospheric species. In planetary atmospheres, atmospheric escape can occur via thermal and nonthermal processes as well as via hydrodynamic and impact erosion processes. The Jeans escape describes the processes of thermal escape (Chamberlain and Hunten 1987). The impact of the solar wind on the upper atmosphere is an example of nonthermal escape. Ionospheric electric field and charge exchange processes can contribute to nonthermal escape. Escape by hydrodynamic and impact erosion are thought to have played an important role in the early Solar System. In planets with weak internal magnetic fields nonthermal processes are a dominant mechanism of atmospheric loss. For example, Mars does not posses an internal magnetic field but has only localized crustal magnetic fields (Acuna et al. 1999), which cannot sufficiently protect the atmosphere from the effects of the solar wind. NASA's MAVEN mission has been investigating the upper atmosphere of Mars and the associated escape processes in an unprecedented detail.

4.8 Ionosphere

The ionosphere consists of the D, E, and F regions, where the E region has been first discovered. The F region has the largest amount of ionization and can be divided into the F_1 and F_2 layers as seen on the right panel of Fig. 4.4. The plasma number density can vary typically from 10^8 to 10^{12} m^{-3}, depending largely on the influence of the solar energy and geomagnetic configuration. These divisions arose from the successive plateaus of electron density n_e observed on records of time delay (virtual height) of radio reflections as the transmitted signal covered a broad range of frequencies.

The overall morphology of the ionospheric regions is governed primarily by the following processes:

- Solar spectrum deposits its energy at various heights depending on the absorption characteristics of the neutral atmosphere,
- Recombination depends on the density,
- Atmospheric composition changes with height,
- Transport processes.

The combination of these processes leads to the formation of the different ionospheric regions with different governing processes.

The basic structure of the ionosphere can be represented by the Chapman[2] production function $P_c = \eta I(\chi, z)\sigma n(z)$, where P_c describes the probability of photon absorption that is proportional to the ionizing photon intensity I as a function of the solar zenith angle χ and altitude z, absorption cross section σ, number density of neutrals $n(z)$, and the ionization efficiency η, which states that not all the incoming energy goes into ionization. Here the I can be considered flux of incident radiation in

---

[2] After Sydney Chapman in 1920s.

$J s^{-1} m^{-2}$. With increasing altitude the photon intensity increases while the number density of neutrals decrease. In other words, as ionizing photons penetrate deeper into the atmosphere, they produce ions as they encounter denser regions. While Chapman production can describe the E region to a good extent, in the F region, where transport plays an important role, the Chapman production function is not a good approximation to the plasma structure.

References

Abdu MA, Ramkumar TK, Batista IS, Brum CGM, Takahashi H, Reinisch BW, Sobral JHA (2006) Planetary wave signatures in the equatorial atmosphere-ionosphere system, and mesosphere-e- and f-region coupling. J Atmos Sol-Terr Phys 68:509–522

Acuna MH, Connerney JEP, Ness NF, Lin R, Mitchell D, Carlson CW, McFadden J, Anderson KA, Reme H, Mazelle C, Vignes D, Wasilewski P, Cloutier P (1999) Global distribution of crustal magnetization discovered by the Mars Global Surveyor MAG/ER experiment. Science 284:790–793

Chamberlain JW, Hunten DM (1987) Theory of planetary atmospheres: an introduction to their physics and chemistry, International geophysics series, vol 36. Academic Press, New York

Jin H, Miyoshi Y, Fujiwara H, Shinagawa H, Terada K, Terada N, Ishii M, Otsuka Y, Saito A (2011) Vertical connection from the tropospheric activities to the ionospheric longitudinal structure simulated by a new earth's whole atmosphere-ionosphere coupled model. J Geophys Res Space Phys 116(A1). doi:10.1029/2010JA015925, http://dx.doi.org/10.1029/2010JA015925

Kallenrode MB (2004) Space physics: an introduction to plasmas and particles in the heliosphere and magnetospheres, 3rd edn. Springer, New york. http://researchbooks.org/3540641262

Laštovička J (2013) Trends in the upper atmosphere and ionosphere: recent progress. J Geophys Res Space Phys 118. doi:10.1002/jgra.50341

Liu H, Miyoshi Y, Miyahara S, Jin H, Fujiwara H, Shinagawa H (2014) Thermal and dynamical changes of the zonal mean state of the thermosphere during the 2009 SSW: GAIA simulations. J Geophys Res Space Phys 119. doi:10.1002/2014JA020222

Pancheva D, Mukhtarov P (2011) Stratospheric warmings: the atmosphere-ionosphere coupling paradigm. J Atmos Sol-Terr Phys 73:1697–1702

Pancheva D, Miyoshi Y, Mukhtarov P, Jin H, Shinagawa H, Fujiwara H (2012) Global response of the ionosphere to atmospheric tides forced from below: comparison between cosmic measurements and simulations by atmosphere-ionosphere coupled model gaia. J Geophys Res Space Phys 117(A7). doi:10.1029/2011JA017452, http://dx.doi.org/10.1029/2011JA017452

Stern DP (2002) A millenium of geomagnetism. Rev Geophys 40. doi:10.1029/2000RG000097

Yiğit E, Medvedev AS (2015) Internal wave coupling processes in Earth's atmosphere. Adv Space Res 55:983–1003. doi:10.1016/j.asr.2014.11.020, http://www.sciencedirect.com/science/article/pii/S0273117714007236

Chapter 5
Waves in Terrestrial and Planetary Atmospheres

A Mechanism of Coupling and Variability

İnsan hakta hak insanda
Ne ararsan var insanda
Çok marifet var insanda
Mademki ben bir insanım

(The human and the truth are one
Whatever you seek, you find in the human
The human consists of insightful knowledge
Because I am a human being)

— Aşık Daimi (20th century)

Abstract First, a brief introduction to wave processes in nature is presented, summarizing basic wave parameters. The main types of atmospheric internal waves, such as gravity waves, solar tides, and Planetary (Rossby) waves, which can propagate upward in planetary atmospheres are described. In the lower atmosphere, observations provide increasing amount of data on the sources of these waves. In particular, small-scale gravity wave sources ought to be studied in more detail. Also, in the upper atmosphere gravity wave signatures are being continuously observed, suggesting a coupling between the lower and upper atmosphere. The effects of internal waves on the terrestrial and planetary atmospheres are then presented, focusing on the small-scale gravity waves generated in the lower atmosphere, in order to link lower atmospheric and upper atmospheric processes. General circulation modeling studies demonstrate that the dynamical and thermal effects of the small-scale gravity waves of lower atmospheric origin substantially contribute to vertical coupling in the atmosphere-ionosphere system.

Keywords Internal waves · Wave parameters · Gravity waves · Tides · Planetary rossby waves · Wave effects · Wave observations · Gravity wave parameterizations · Vertical coupling

5.1 Notion of Waves

Waves are fundamental processes in the universe. They are characterized by a broad range of spatial and temporal scales and are encountered in physical processes involv-

© The Author(s) 2015   53
E. Yiğit, *Atmospheric and Space Sciences: Neutral Atmospheres*,
SpringerBriefs in Earth Sciences, DOI 10.1007/978-3-319-21581-5_5

ing microscales, e.g., in quantum mechanics, as well as in macroscale processes that are seen, for example, in geophysics and astronomy. Depending on the scientific context, they are sometimes termed as "oscillations", "disturbances", "fluctuations", or "vibrations", even sometimes simply as "variations". These expressions will be used interchangeably, depending on the context.

Although the notion of waves sounds like an abstract concept, waves are a part of nature and can be observed pretty much everywhere around us. The visible light that allows us to see our environments is a wave. With a wavelength range of 380–750 nm (i.e., $7.9 - 4.0 \times 10^{14}$ Hz), the visible light belongs to the spectrum of electromagnetic waves, which represent a road range of waves from radio waves to γ-rays, having a 15 order of magnitude scale difference (i.e., $10 - 10^{-14}$ m).

While you are outside in nature and observing the sky during a sunset, it is often possible to see the signatures of atmospheric waves in clouds, if they are present. Figure 5.1 shows a picture that I have taken in Fairfax, Virginia, USA, before a sunset. In the red-shining clouds, internal gravity wave structures can be seen. If you are walking on the beach, you will encounter the most familiar form of waves. Namely, you can observe water waves, which are surface waves, propagate toward the coast. Seismic waves can be detected during and after earthquakes. When you throw a pebble in a pool, you will see concentric ripples on the surface of the water

Erdal Yiğit 11/19/2013

Fig. 5.1 The sky before sunset in Fairfax Virginia on 19 November 2013, demonstrating wave-like signatures in clouds

propagating radially outward. A snake produces waves that travel from its head to tail, which stay stationary with respect to the ground, so the snake moves forward. Every time you play a song, for example, with a stringed instrument, such as a guitar or a bağlama/saz, you are generating a sequence of musical notes by vibrating the string in various ways.

Wave motion, in general, implies that a certain physical property is transmitted from one place to another by means of a medium, where the medium itself is not transported. All fundamental states of matter, liquid, solid, gas, and plasma, can carry waves. Therefore, waves are an intrinsic property of our nature.

In the context of atmospheric and space research, the observed rapid time-varying changes of field variables are often described as "wave-like" variations. In particular, this description occurs when the nature of the observed variability is not well-defined or not well-known. Especially, in terrestrial and planetary atmospheres, the dynamical roles of waves and their significance for global atmospheric circulation are increasingly appreciated because of the ability of waves to transfer energy and momentum, and transport chemical species in spacetime ($\mathbf{r}$, t). Thus, waves can ultimately alter the characteristics of the medium they propagate in by causing coupling within the different regions of the same medium. It is thus necessary to better understand the concept of waves, how they are generated, what their propagation and dissipation characteristics are, and what the resultant effects on the medium, i.e., the atmosphere, are. The processes of generation, propagation and dissipation, and the resulting effects are fundamental processes that underline every wave process in nature. We will primarily focus on the aspects of internal wave observations and modeling the propagation and dissipation of internal waves in the atmosphere.

This chapter will summarize some basic concepts of wave physics (Sect. 5.2), stating the wave equation (Sect. 5.3), present some types of internal waves encountered in terrestrial and planetary atmospheres (Sect. 5.4) and will review some recent research in the observation and modeling of the effects of internal gravity waves in the atmosphere (Sects. 5.9 and 5.10).

5.2 Introduction to Basic Wave Parameters

Spatial and temporal scales (characteristics) of waves are described by the wavelength λ and period τ, respectively. The period is the amount of time required to perform one complete oscillation and is the inverse of frequency f

$$\tau = \frac{1}{f}, \quad [\text{s}] \tag{5.1}$$

from which it can be concluded that frequency is the number of oscillations per unit time, measured typically in seconds. A larger/smaller number of oscillation per unit time means larger/smaller frequency. Accordingly, high-frequency waves have small periods and low-frequency waves have large periods.

Angular frequency ω is the frequency associated with a complete wave cycle and is given by

$$\omega = 2\pi f = \frac{2\pi}{\tau}. \quad [\text{rad s}^{-1}] \tag{5.2}$$

The spatial scale is given by the wavelength λ, which is the distance over which one oscillation is completed.

$$\lambda = \frac{2\pi}{K}, \tag{5.3}$$

where $K = |\mathbf{k}| = \sqrt{k^2 + l^2 + m^2}$ is the magnitude of the wave vector $\mathbf{k}$, described in terms of the zonal (k), meridional (l), and the vertical wavenumbers (m)

$$\mathbf{k} = (k, l, m), \tag{5.4}$$

which is often expressed in terms of the Cartesian coordinates as $k_x = k$, $k_y = l$, and $k_z = m$. Using Eq. (5.3) the total wavelength λ can be expressed in terms of the horizontal, meridional, and vertical wavelengths

$$\lambda = \left(\frac{1}{\lambda_x^2} + \frac{1}{\lambda_y^2} + \frac{1}{\lambda_z^2} \right)^{-2}. \tag{5.5}$$

Often in atmospheric physics, we define a horizontal wavelength λ_H in terms of a horizontal wavenumber as $\lambda_H = 2\pi/k_H$ with $k_H = \sqrt{k^2 + l^2}$. Figure 5.2 illustrates a simple one-dimensional sinusoidal wave $f(x) = \sin x$ along the x-axis with the wavelength $\lambda = \lambda_x$ and the amplitude A, which is the maximum departure from the equilibrium position.

Fig. 5.2 Illustration of the parameters of a typical sinusoidal wave (f(x) = sin x) (*blue*) in one dimension from 0 to 2π. λ is the wavelength and A is the amplitude

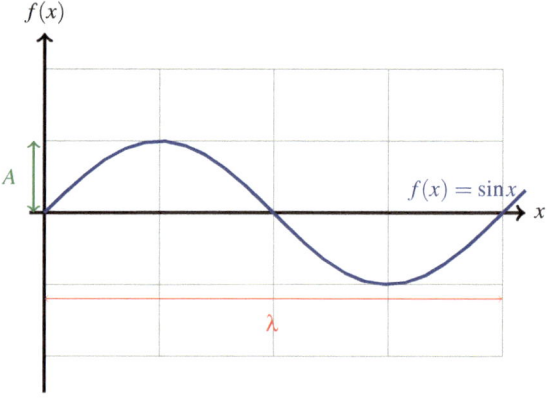

Further kinematic characteristics of waves can be illustrated by a plane wave, which is a general solution to the equation of motion of simple linear systems or small-amplitude waves. We can then express a field variable, such as the pressure p, using a plane wave representation.

$$p(\mathbf{r}, t) = \Re\{p_0 \exp[i\,(\mathbf{k} \cdot \mathbf{r} - \omega t]\}, \tag{5.6}$$

where $\mathbf{r} = (r_1, r_2, r_3) = (x, y, z)$ is the position vector; t is time; ωt describes the temporal characteristics of the wave and

$$\mathbf{k} \cdot \mathbf{r} = \sum_i k_i r_i. \tag{5.7}$$

is the spatial extent of the wave. Equation (5.6) describes a *progressive wave*, where the real part of the expression has the actual physical meaning. In three-dimensional space

$$\mathbf{k} \cdot \mathbf{r} = kx + ly + mz. \tag{5.8}$$

The spatiotemporal variable phase of a plane wave is described by its phase as

$$\phi(\mathbf{r}, t) = \mathbf{k} \cdot \mathbf{r} - \omega t. \tag{5.9}$$

On a given plane, the phase is constant (C):

$$\phi(\mathbf{r}, t) = \mathbf{k} \cdot \mathbf{r} - \omega t = C \tag{5.10}$$

In two dimensions, this yields

$$kx + mz - \omega t = C \tag{5.11}$$

Note that the negative local rate of change of the phase yields the angular frequency ω and the spatial gradient of the phase describes the wave vector $\mathbf{k}$.

$$-\frac{\partial \phi}{\partial t} = \omega \qquad \frac{\partial \phi}{\partial \mathbf{r}} = \mathbf{k}. \tag{5.12}$$

The speed of the propagation of a given phase is described by the phase speed c

$$c = \frac{\omega}{K} = \left|\frac{\partial \phi}{\partial t}\right| \times \left|\frac{\partial \phi}{\partial \mathbf{r}}\right|^{-1} \tag{5.13}$$

The relation between the wavenumber and the angular frequency yields the dispersion relation of propagating waves, i.e., in general, $\omega(K)$. A dispersive wave means that the various sinusoidal components of a disturbance originating at a given location and time are found in different places at a later time. Energy is then said to be dispersed. Acoustic (sound) waves are nondispersive, meaning that they preserve their shape. The group velocity $\mathbf{c}_g$ describes the direction of energy flux caused by the wave and is given by

$$\mathbf{c}_g = \left(\frac{\partial \omega}{\partial k}, \frac{\partial \omega}{\partial l}, \frac{\partial \omega}{\partial m} \right) \tag{5.14}$$

Waves in fluids take place because of the action of a restoring force on the fluid parcels that have been displaced from their equilibrium positions. The restoring force can comprise compressibility, gravity, rotation, or electromagnetic effects. In a realistic atmosphere, the fluid is in motion due to the continuous circulation of the atmosphere and the wave motion is essentially embedded in a moving fluid. Then the intrinsic phase speed $\hat{c}$ of a wave along the x-direction is

$$\hat{c} = c - u. \tag{5.15}$$

Because of the relative motion of the wave and the fluid in the atmosphere, the wave frequency is Doppler-shifted, which is the intrinsic wave (angular) frequency $\hat{\omega}$

$$\hat{\omega} = (c - u)k = \omega - uk. \tag{5.16}$$

When the wave propagates in the same direction as the flow, $\hat{\omega}$ is lower than the case when the wave propagates in the opposite sense to the flow. The intrinsic parameters are fundamental for the description of wave-mean flow interactions.

5.3 Wave Equation

The equation that describes a wave propagation in space-time is called the wave equation. There are a number of forms of wave equations depending on the type of wave. One simplified customary form of the wave equation in one dimension is

$$\frac{\partial^2 \psi(x, t)}{\partial t^2} = c^2 \frac{\partial^2 \psi(x, t)}{\partial x^2}, \tag{5.17}$$

where $\psi(x, t)$ is the wave function and the differential equation is a second-order time-dependent conservative equation. The above equation can readily be expanded to three dimensions to describe the evolution of $\psi(\mathbf{r}, t)$. A plane wave $\propto \exp[i(\mathbf{k} \cdot \mathbf{r} - \omega t)]$ is a general solution to the wave equation.

5.4 Internal Waves

The atmosphere can be understood as an ideal geophysical fluid that is pervaded by waves of various spatio-temporal scales. The term "internal" describes the ability of waves to propagate "internally", that is, vertically upward within the atmosphere, unlike "external" modes, in which all layers oscillate in sync, while the harmonics propagate horizontally. Real physical systems in nature are capable of oscillating in many different frequencies. Various characteristic vibrations of a system are called *modes* of that system. As a typical many-particle system, the atmosphere possesses many modes of vibrations. Overall, internal waves exist because Earth's atmosphere is overall stably stratified (see Sect. 2.12). Internal waves have great dynamical significance for the atmosphere because of their ability to propagate upward and interact with the flow (Sect. 5.10). This upward propagation implies that they can reach higher altitudes in the atmosphere, providing a coupling mechanism. As they propagate upward, their intrinsic properties continuously change due to changes in the background atmosphere. Coupling in this context means that energy or momentum has been transferred from one region to another region in the atmosphere or certain chemical species have been transported.

One common way of classifying atmospheric waves is with respect to their spatial scales. Owing to a variety of generation sources in the lower atmosphere, Earth's atmosphere possesses a broad spectrum of waves ranging from very small- (e.g., small-scale gravity waves) to planetary-scale waves (e.g., tides, Rossby waves, Kelvin waves), as summarized in Table 5.1. Overall, internal wave periods vary from few minutes to tens of days. They also cover a large range of spatial scales with horizontal scales of 10 km to planetary scales. While small-scale GWs have typical horizontal wavelengths λ_H of several km to several hundred km, horizontal scales of thermal tides and planetary waves are comparable to the circumference of Earth. Such waves are characteristic of all planetary atmospheres that are able to sustain a stably stratified neutral gas. The temporal and spatial scales of these waves can vary from planet to planet. The size of a given planet will influence the scale of the waves.

Table 5.1 Typical temporal and spatial scales of internal waves in the terrestrial atmosphere

Internal wave	Temporal scales	Horizontal scale
Gravity wave	Minutes to several hours $(2\pi/f, \; f = 2\Omega \sin\phi)$	10–1000 km
Solar tide	1, 1/2, 1/3 days	Planetary scale
Planetary wave	2 to few tens of days	Planetary scale
Kelvin wave	3–20 days	Planetary scale

Adopted from Yiğit and Medvedev (Table 1, 2015)

5.5 Wave Generation

Although different processes can generate various types of internal waves, in general, the lower atmosphere, where the weather takes place, is the primary source of these waves. This is why internal waves originating from the lower atmosphere are said to be primary waves. For example, meteorological processes, such as convection (Song et al. 2003; Beres et al. 2004; Song et al. 2007), frontogenesis (Gall et al. 1988), nonlinear interactions (Medvedev and Gavrilov 1995), and thunderstorms (Curry and Murty 1973) can generate a broad spectrum of gravity waves. More details of gravity wave sources can be found in the work by Fritts and Alexander (2003). Solar tides can be generated in the lower atmosphere by tropospheric latent heat release (Hagan and Forbes 2002, 2003). Internal waves can propagate upward and grow in amplitude due to exponentially decreasing neutral mass density ρ (in order to satisfy wave action conservation). The wave disturbances can have small amplitudes at their sources in the lower atmosphere, but with increasing altitude, the amplitudes can become significant at higher altitudes in the thermosphere, and are all subject to various dissipation processes. This dissipation is the mechanism of wave mixing and transfer of momentum and energy from waves to the mean flow. Primary gravity wave-mean flow interactions can then lead to secondary wave generation at higher altitudes (Vadas et al. 2003). I next summarize some of the common internal waves encountered in terrestrial and planetary atmospheres, focusing on GWs.

5.6 Gravity Waves

Internal gravity waves are fundamental features of any stably stratified planetary atmosphere. This property implies that an infinitesimally small air parcel that is departed from its equilibrium position z_{eq} by some external disturbances can return to its original equilibrium position. Then the fluid parcel is said to oscillate around this equilibrium position. The associated oscillations are buoyancy oscillations as atmospheric buoyancy acts as the restoring force. Therefore, gravity waves are often called buoyancy waves.

Figure 5.3 illustrates the vertical distance δz a fluid parcel covers between z_1 and z_2. The upward/downward movement can occur with an angle of β with respect to the vertical direction, which is represented by the distance δs. The distances δz and δs are related as

$$\delta z = \delta s \, \cos \beta. \tag{5.18}$$

The vertical buoyancy force per unit mass is then given in terms of the buoyancy frequency (or Brunt-Väisälä frequency) N as $-N^2 \delta z$ and along the tilted path it is $-N^2 \delta z \cos \beta = -N^2 \delta s \cos^2 \beta$. Then, the equation of motion for a particle along δs is

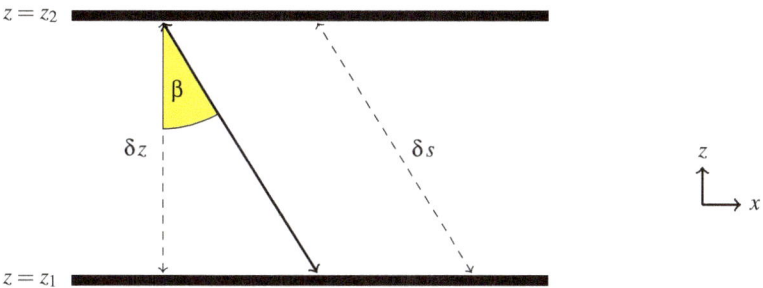

Fig. 5.3 Illustration of a gravity wave oscillation in two dimensions between two altitude levels z_1 and z_2 separated by a vertical distance of δz

$$\frac{\mathrm{d}^2(\delta s)}{\mathrm{d}t^2} = -N^2 \cos^2 \beta, \tag{5.19}$$

which has a general solution of

$$\delta s = \exp[\pm i (N \cos \beta)t], \tag{5.20}$$

describing a harmonic oscillation with the angular frequency $\omega = N \cos \beta$. Maximum frequency is given by the buoyancy frequency, i.e., $\omega = N$, for the scenario of a perfect vertical oscillation. The oscillation of an air parcel around its equilibrium position is analogous to the simple harmonic oscillator. A detailed mathematical treatment of buoyancy oscillations can be found in the seminal books by Gossard and Hooke (1975) and Nappo (2002).

5.6.1 Boussinesq Approximation

If the vertical displacement δz of an air parcel is relatively small, then the associated density change $\delta \rho$ can be assumed to be small as well. If the mass density is expressed in terms of a mean (background) value and a fluctuating part, i.e., $\rho = \bar{\rho} + \rho'$, where $\rho' = \delta \rho$, then one can make the approximation that density has a constant mean value $\rho(z) = \bar{\rho}$, except in the buoyancy term in the vertical momentum equation. Therefore, density fluctuations are considered only in combination with gravity g. In a compressible atmosphere, fluctuations are complicated because the mass density is a function of temperature and pressure, $\rho(p, T)$, thus fluctuations in density would imply variations in temperature and pressure. With the Boussinesq approximation dynamical equations can greatly be simplified. For example, the application of this approximation in GW dynamics yields a simplified form of the continuity equation

$$\nabla \cdot \mathbf{u} = 0, \tag{5.21}$$

where incompressibility is assumed and the flow is divergence-free. Also, the density fluctuations are assumed to be much smaller than the constant background value, $|\rho'/\bar{\rho}| \ll 1$, which is satisfied if and only if the vertical scale of the variations are much smaller than the scale height, i.e., $\lambda_z \ll H$. Overall, the dynamical consequence of the Boussinesq approximation is that the fluctuating changes in density due to local pressure variations are negligible; the fluid can be treated as incompressible; and acoustic waves are eliminated, as $\rho' = \delta\rho$ and $p' = \delta p$ are separated.

5.6.2 Gravity Wave Dispersion Relation

How the temporal and spatial characteristics of GWs are related to each other is described by the dispersion relation $\omega(K)$. From Eqs. (5.2)–(5.3) we get for the horizontal phase speed c_h

$$c_h = \frac{\omega}{k} = \frac{\lambda_x}{\tau}. \tag{5.22}$$

Therefore, for a given wave period larger scale waves propagate faster than smaller scale waves and we get wave broadening, i.e., dispersion. For the midfrequency GWs ($N \gg \hat{\omega} \gg f$), we have

$$\hat{\omega} = \left|\frac{k_H}{m}\right| N. \tag{5.23}$$

5.7 Large-Scale Waves

Thermal tides (solar tides) are large-scale internal waves that are generated by the periodic heating of the solar radiation absorption in the atmosphere. The diurnal tide is the most known tide with a period of $\tau = 24$ h and angular frequency of $\omega = 2\pi/(24\,\text{h})$. However, due the nonlinear nature of the atmosphere, higher frequency tidal oscillations are produced as well, with frequencies $\omega_n = n\Omega$, where $n = 1, 2, 3$, etc., denote the higher-order harmonics (Forbes 1984). Tidal oscillations can be represented by the Fourier series, which allows every periodic motion to be represented by a superposition of trigonometric functions. For tides, we have

$$\psi_{ns} = A_{ns}(n\Omega t + s\lambda + \phi_{ns}), \tag{5.24}$$

where A_{ns} and ϕ_{ns} are the amplitude and the phase of the "ns" harmonic, λ is the longitude, $s = 0, \pm 1, \pm 2, \ldots$ is the zonal wavenumber, respectively. Positive and negative zonal wavenumbers give westward and eastward propagating harmonics.

For $s = n$ we have a Sun-synchronous tide propagating westward with the apparent motion of the Sun. Such tides are called migrating solar tides. Non-migrating tides can propagate westward and eastward and are represented by $s \neq n$.

5.8 Wave Decay

Basic concepts of the free vibrations of physical systems can be illustrated by a harmonic oscillator composed of a single object of mass m attached to a spring with a linear elastic property. All the mass of the system is then assumed to be concentrated in the object. Then the well-known Hooke's law

$$\ddot{x} = -\frac{k}{m}x, \tag{5.25}$$

where k is the spring constant and $\omega^2 = k/m$ is the angular frequency, describes the associated harmonic oscillations. The total mechanical energy is conserved and is composed of the potential energy $kx^2/2$ plus the kinetic energy $mu^2/2$. Such an oscillation can be solved with a complex exponential ansatz $\propto \exp[i(kx - \omega t)]$. Basic notion of an undamped simple harmonic oscillator does not include decay processes. In a realistic world, oscillations are always subject to damping. Then one speaks of wave dissipation. Imagine a tune produced by an instrument. Without damping, the tune would continue indefinitely. In reality, it decays. We can represent the decay processes by introducing a damping term in Eq. (5.25) as

$$\ddot{x} = -\frac{k}{m}x - \frac{b}{m}\dot{x}, \tag{5.26}$$

where $\omega_0^2 = k/m$ is the undamped angular frequency of the system and $\beta = b/m$ is the damping factor (with a dimension of frequency). The damping is assumed to be linearly proportional to the motion, for example, the type of damping an object would experience moving in a fluid. So, what are the consequences of this damping to the free vibrations of the system? The primary effect of the damping is to modify the angular frequency and reduce the oscillation amplitude A. As the total mechanical energy of a simple harmonic oscillator is $kA^2/2$, the energy decay is exponential. In principle, if the system is highly nonlinear, the associated wave damping processes can be nonlinear as well.

In planetary atmospheres, we do not simply have a spectrum of wave modes that freely oscillate. Waves are subject to various dissipation mechanisms. Internal waves can propagate upward in stably stratified atmospheres and are subject to dissipation effects whose intensity depend on the characteristics of the background atmosphere. Wave dissipation is an important internal mechanism of planetary atmospheres and can generate natural variability. Via damping, the mechanical energy of the wave is dissipated and shared with the flow. As internal waves can propagate over large distances away from their generation regions, they can carry momentum and energy away from their source and are thus a prime candidate for causing atmospheric coupling. As the vertical structure of the atmosphere-ionosphere system strongly varies, the associated propagation and, in particular, wave dissipative processes vary significantly with altitude as well as with latitude and longitude. To obtain a precise picture of internal wave dissipation, all physical processes influencing a system ought to be adequately represented in the equations of atmospheric dynamics (Sect. 3.9). The resulting variations of the wave parameters, such as the frequency and the amplitudes, then represent the reality better. In Earth's atmosphere damping due to nonlinear diffusion (Weinstock 1976, 1982; Medvedev and Klaassen 1995), ion-drag, molecular diffusion and thermal conduction (Vadas and Fritts 2005), eddy diffusion, and radiative damping. For the first time, the extended nonlinear gravity wave parameterization of Yiğit et al. (2008) has accounted for the superposition of all these wave dissipation effects to estimate more accurately gravity wave effects in Earth's whole atmosphere system. The wave model can run stand-alone as well as within a general circulation model. The use of this scheme in determining the effects of gravity waves in the circulation of the atmosphere will be discussed later (Sect. 5.10). This scheme has also successfully been applied in Mars atmosphere (e.g., Medvedev et al. 2011a, 2013).

Earlier studies have focused on the effects of gravity waves on the general circulation of the middle atmosphere (e.g., Holton 1982; Garcia and Solomon 1985; Hunt 1986; Hines 1991). Some earlier studies have also focused on wave dissipation in the thermosphere (e.g., Pitteway and Hines 1963; Hines and Hooke 1970; Klostermeyer 1972; Hickey and Cole 1988).

In the next section, I will present a brief review of the observations of internal wave signatures in terrestrial and planetary atmospheres.

5.9 Observation of Waves in the Atmosphere

If internal waves would not propagate upward away from their sources and interact with the atmosphere, then no wave signature would ever be detected by observations. However, wave signatures in the atmosphere are quiet common. I will primarily focus on the recent observations of internal gravity waves in terrestrial and planetary atmospheres. Probably, due to its broad range of temporal and spatial scales, gravity wave observations are among the most challenging observational activities.

There is a variety of observational techniques, such as, lidar, radar, balloon, and satellite-born measurements to detect wave signatures in the atmosphere. In particular, global observations with satellites are a popular research tool. Table 5.2 lists some of the research-oriented satellite missions that provide data on the current state of

Table 5.2 Characteristics of CHAMP (CHAllenging Minisatellite Payload), GRACE (Gravity Recovery and Climate Experiment), TIMED (Thermosphere Ionosphere Mesosphere Energetics and Dynamics), DMSP (Defense Meteorological Satellite Program), and Swarm satellites in the context of atmospheric observations

Satellite	Instrument	Data	Altitude (km)
CHAMP	Ion drift meter	Ion density	~456–310
DMSP		Ion density	~850
GRACE	GRACE-A/B	TEC	~490–440
TIMED	GUVI	O/N_2	~630
	TIDI	Wind, temperature	
	SABER	Temperature	
SWARM	VFM	Vector magnetic field	~450, 550

Some selected instruments that measure relevant upper atmosphere parameters are listed in the second coulumn along with the data product in the third column. The initial altitude at launch and the altitude the satellite decayed to are listed as well

the atmosphere-ionosphere system. Some of the instruments on board these satellites are listed as well along with some measured parameters and the approximate altitude of the satellite. A variety of upper atmosphere parameters, such as, ion densities, total electron content (TEC), O/N_2 ratio, magnetic field strength, neutral winds and temperature can be globally observed by various satellites operating between 400 and 850 km. European Space Agency's (ESA) Swarm mission is one of the recent satellite missions that is dedicated to surveying the upper atmosphere and Earth's magnetic field. It is expected to provide an unprecedented view of the geomagnetic field and its spatiotemporal evolution. I next present a brief overview of observations of internal waves in the lower and upper atmosphere.

5.9.1 Observations in the Lower Atmosphere

Internal waves are generated by a variety of sources in the lower atmosphere and observations can help characterize the distribution of the associated wave activity close to their sources. Due to the nonlinearity of generation processes, we expect a complex global structure of wave activity.

Long-duration balloon measurements can be conducted to determine the GW activity in the lower atmosphere. Figure 5.4 shows the average distribution of GW horizontal density-weighted momentum fluxes $\rho_0 u' w'$ in the Southern Hemisphere as evaluated by Hertzog et al. (2008) during the Vorcore Campaign in Antarctica. These results indicate that the GW activity is distributed very heterogeneously in the Southern Hemisphere high-latitudes. Largest fluxes are located around the Antarctic peninsula, peaking with 30 mPa. Such balloon campaigns can include up to 2500 observations in each box seen in the figure. Hertzog et al. (2008)'s measurements

Fig. 5.4 Absolute
momentum flux in mPa
retrieved from long-duration
balloon measurements. The
boxes are $10° \times 5°$ longitude
$\times$ latitude resolution.
Dashed lines indicate region
of observations with less
than 200 observations, which
have not been statistically
analyzed. Adopted from the
work by Hertzog et al.
(2008)

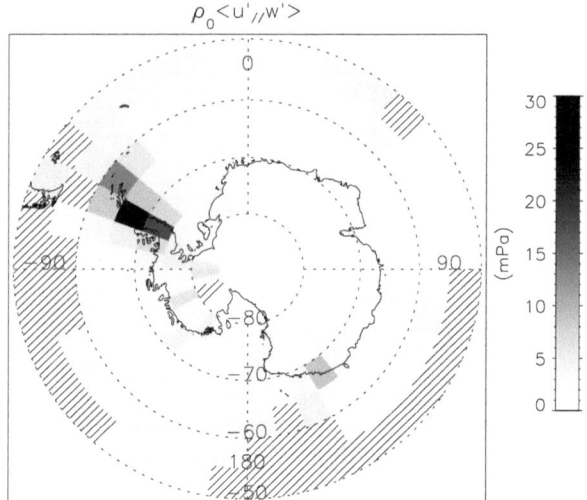

highlighted the significance of this peninsula region in the generation of gravity waves as were suggested by a number of researchers (Hoffmann et al. 2013).

While measurements with balloons or ground-based instruments provide high-quality gravity wave data, they have a limited global coverage. Satellites provide a near-global view of atmospheric fields and wave activity. Figure 5.5 presents the High Resolution Dynamics Limb Sounder ((HIRDLS), Gille et al. 2008) observations of the horizontal distribution of gravity wave momentum flux at 25 km in the Northern and Southern Hemispheres averaged over the associated summertime periods (Fig. 1, Ern and Preusse 2012). Significant longitudinal variability and hemispheric differences are seen in the total momentum fluxes. There is a distinct annual variation.

5.9.2 Observations in the Upper Atmosphere

Oscillations in the upper atmosphere have routinely been observed already in 1950s, although the nature of these variations were not clear. For example, Munro (1950) detected ionospheric disturbance as changes in the virtual height of ionospheric reflection at a fixed radio frequency. The disturbances were short-period and horizontally propagating. Hines (1960) has discussed the theoretical basis for the presence of atmospheric gravity waves at ionospheric heights. With simultaneous pairs of clouds photographs, Witt (1962) has detected wave motion in the clouds and argued that it probably is associated with GWs.

Recently, GW signatures have been frequently observed in the thermosphere and ionosphere by a number of authors (Djuth et al. 2004; Livneh et al. 2007; Djuth et al. 2010; Vadas and Crowley 2010; Shume et al. 2014). Satellites can provide an

Fig. 5.5 A global view of the average total gravity wave momentum flux in the **a** Northern Hemisphere and **b** Southern Hemisphere at 25 km altitude during the summer time in both hemispheres, retrieved from the High Resolution Dynamics Limb Sounder (HIRDLS) observations. Adopted from (Fig. 1, Ern and Preusse 2012)

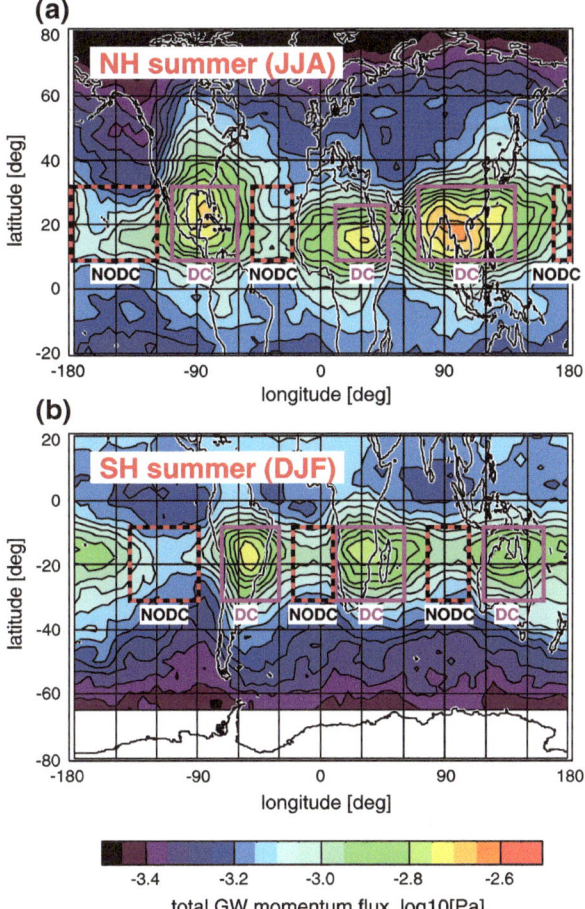

unprecedented global view of GW activity in the upper atmosphere, as well. Park et al. (2014) have used the CHAMP satellite data in order to determine the global maps of GW activity in the upper thermosphere (Fig. 5.6). They have analyzed the dayside data (09–18 LT) for solar minimum conditions, focusing on the low-latitude and midlatitude regions. The highest GW activity are seen at high-latitudes, in particular, during June solstice, the Antarctic peninsula region has the highest activity. GWs demonstrate higher longitudinal variability in the Southern Hemisphere than in the Northern Hemisphere. Park et al. (2014) argue that their observations are consistent with Yiğit and Medvedev (2010)'s GCM simulations of the solar cycle variations of thermospheric GW activity.

Fig. 5.6 A global view of average thermospheric gravity wave activity in terms of relative mass density perturbations at equinox, June solstice and December solstice retrieved from the CHAMP satellite for solar minimum conditions (2006–2010). The color bar is scaled logarithmically and gray regions represents the lowest level of the color scale. The *dashed line* shows the geomagnetic latitude. (Fig. 2, Park et al. 2014)

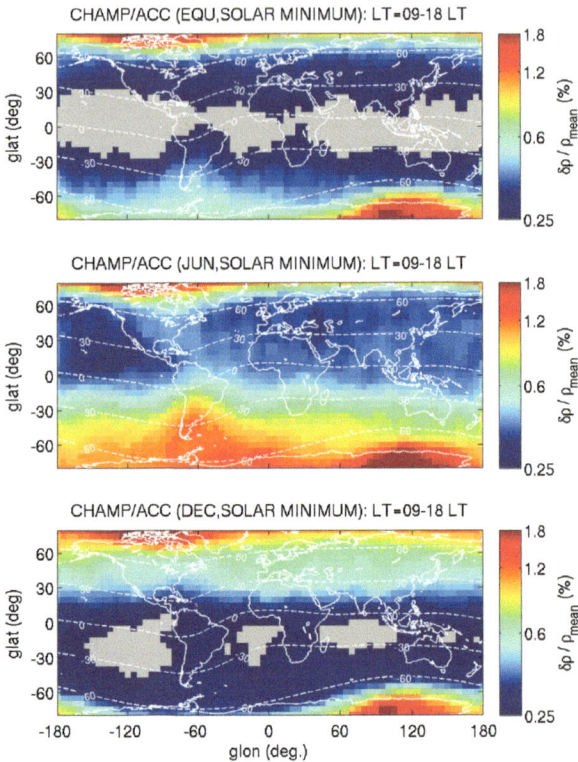

5.10 Effects of Waves in Terrestrial and Planetary Atmospheres

Internal waves greatly shape the dynamical and thermal structure of terrestrial and planetary atmospheres. I will primarily focus on some recent developments in quantifying the effects of internal waves in Earth's atmosphere.

The expression of the "effects of waves" implies that (1) internal waves interact with the atmosphere and (2) there is some quantitative change in motion, temperature and/or concentration, which are described by the conservation equations. Although, observations provide an evidence of the presence of wave signatures and of the associated perturbations in planetary atmospheres, it is not trivial to quantify the actual link between atmospheric circulation and wave effects. General circulation models are excellent tools for the simultaneous quantification of internal wave effects and the resulting changes in the general circulation of the atmosphere. Concerning wave effects, I will mostly focus on the modeling aspect of these waves. A more comprehensive and recent review of internal wave effects in Earth's atmosphere can be found in the work by Yiğit and Medvedev (2015).

5.10.1 Gravity Wave Processes

Internal gravity wave propagation and dissipation greatly shape the dynamical structure of the middle atmosphere (e.g., Holton 1983; Fritts 1984; Garcia and Solomon 1985; Hunt 1986; Fritts and Alexander 2003; Fritts et al. 2006; Becker 2011; Smith 2012). Gravity waves are largely responsible for the departure of the middle atmosphere form a radiative equilibrium as they dynamically reverse the meridional temperature gradient around the mesosphere and the lower thermosphere. The coldest spot on Earth is found in the summer mesopause where temperatures can fall down to 120 K. In the mesosphere and thermosphere gravity waves substantially shape the dynamical and thermal structure of the atmosphere. Because of the relatively small scales of gravity waves, not all parts of atmospheric gravity waves can self-consistently be resolved in GCMs. Therefore, parameterizations are used, which next section discusses.

5.10.1.1 Gravity Wave Parameterizations in Models

Historically, the departure of atmospheric energy budget from radiative equilibrium has revealed the presence of significant wave-induced forcing of the middle atmosphere (Leovy 1964). As the source of this wave drag was not clearly known at that time, a so-called *Rayleigh drag*, an artificial linear friction force, which somewhat improved model simulations, has been used (e.g., Schoeberl and Strobel 1978; Holton and Wehrbein 1980). However, this artificial drag, being proportional to the background wind, could not reproduce the observed wind reversals in the MLT, but could only close MLT winds (i.e., it reduced the wind speeds to zero). The Rayleigh drag is given by

$$a_r = -k_r \, \bar{u}, \tag{5.27}$$

where k_r is a proportionality constant. In principle, a simple Rayleigh drag is physically unrealistic, thus the necessecity of a more realistic representation of wave drag has been accepted in the form of GW parameterisations (Lindzen 1981).

   Atmospheric general circulation models (GCM) cannot resolve the entire spectrum of gravity waves due to their limitation in spatial and temporal resolution. Large-scale portion is resolved to some extent, but the small-scale portion is largely not captured in models. Therefore, parameterization of gravity waves are used in most global-scale models in order to represent the missing portion of small-scale gravity waves (e.g., Garcia et al. 2007; Yiğit et al. 2009). For this, using typically an empirical distribution of gravity wave fluxes at a source level, the upward propagation and dissipation of the individual waves in the spectrum are quantified in each horizontal grid point and altitude level in the model. The principle of quantifying the wave effects are independent of the model grid structure.

The quantified effects are then included as additional terms in the governing equations of atmospheric dynamics (Sect. 3.9) to better estimate the transport of energy, momentum, and mass within the atmosphere. For example, in the model integration of the (zonal) wind tendencies, we have a resolved portion (a) and an unresolved component associated with small-scale GWs (a_{gw}) due to the limited resolution

$$\frac{du}{dt} = \underbrace{a}_{resolved} + \underbrace{a_{gw}}_{unresolved}. \tag{5.28}$$

Therefore, a "good" gravity wave parameterization should represent this unresolved portion adequately. Using improved gravity wave parameterizations in GCMs yields better comparison with observations.

A number of nonorographic gravity wave parameterizations have been developed (e.g., Lindzen 1981; Holton 1982; Matsuno 1982; Medvedev and Klaassen 1995; Hines 1997a, b; Alexander and Dunkerton 1999; Warner and McIntyre 2001; Yiğit et al. 2008), which quantify the effects of GWs produced by various sources in the lower atmosphere. The orographic GW parameterizations are designed to explicitly represent the effects of topography on the general circulation of the atmosphere (e.g., Palmer et al. 1986; McFarlane 1987; Lott and Miller 1997). Convective GW parameterizations approximate GW effects produced by convective activity in the lower atmosphere (e.g., Chun and Baik 2002; Beres et al. 2004; Schirber and Alexander 2014). This process is thought to produce a broad spectrum of waves that can propagate to high altitudes in the atmosphere and produce appreciable effects (Vadas and Liu 2009). Technically, it is possible to include the convectively generated GWs within the framework of nonorographic parameterizations. A comprehensive review of earlier developments of gravity wave parameterizations in numerical weather prediction models can be found in the paper by Kim et al. (2003).

5.10.1.2 Initial Gravity Wave Spectrum

A wave spectrum typically includes discrete harmonics and the associated activities of these waves. Thus, it provides information on wave characteristics and how the properties are distributed among the individual harmonics assumed to be present in the spectrum. This distribution could be a collection of characteristic wavelengths (wavenumbers). The spectral description of GWs requires a definition of the shape of the spectrum and knowledge of what parameters control wave characteristics. For the first time, VanZandt (1982) proposed that the spectrum of observed mesoscale fluctuations in the atmosphere is independent of climatological conditions, geographical location, and season. This implies that the power spectral density (PSD) of mesoscale horizontal velocity fluctuations associated with waves versus frequency, horizontal and vertical wavenumber have a universal relationship. The so-called universality of the atmospheric wave spectrum has first been discovered by Dewan and Good (1986).

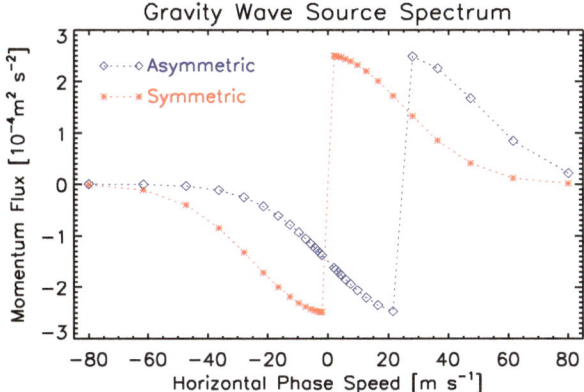

Fig. 5.7 Gravity wave horizontal momentum flux spectrum at a source level ($\sim$15 km) in the lower atmosphere as it is specified within the spectral nonlinear gravity wave parameterization Yiğit et al. (2008) and implemented into the general circulation model. The *asterisks*, *triangles*, and *rectangles* denote the fluxes for individual harmonics used in various simulation experiments. The maximum momentum flux is 0.00025 m^2 s^{-2} and the spectrum width is $c_w = 35$ m s^{-1}. The connecting *dotted lines* highlight the Gaussian shape of the spectrum. An asymmetric spectrum is represented by $u_0 = 20$ m s^{-1}. Adopted from Yiğit et al. (2009, Fig. 1)

Gravity wave parameterization all require a specification of an initial gravity wave activity at an appropriate source level. In global models, this level is numerically chosen to be either close to the lower boundary or around the tropopause. The middle atmosphere is quite sensitive to the spectral characteristics and to the choice of spectrum launch level (Manzini and McFarlane 1998). Figure 5.7 illustrates a GW source spectrum used by the extended nonlinear GW parameterization of Yiğit et al. (2008) in the Coupled Middle Atmosphere Thermosphere-2 GCM. The spectrum is expressed in terms of wave momentum fluxes $\overline{u'w'}_i$ as a function of horizontal phase speeds c_i of the ith harmonic. The spectrum contains typically 30 harmonics that are composed of positive and negative phase speeds. It has a distinct Gaussian shape with peak fluxes around slow waves (low horizontal phase speeds). Earlier modeling studies have provided substantial insight into GW dynamics in the middle atmosphere with relatively simple GW spectra (Garcia and Solomon 1985). Gravity wave generation in the lower atmosphere is a spatially and temporally variable process. The GW spectra used in GCMs account for a statistical representation of GW activity, that is, they reflect an average characteristics of GW processes. GW parameterizations have indeed various limitations. They all assume GWs propagate only vertically and the response of the resolved fields are instantaneous. Nevertheless, even observations of GW activity have to assume some averaging in form of binning in order to produce a distribution wave activity (Hertzog et al. 2008).

5.10.1.3 Gravity Wave Effects on the General Circulation of the Atmosphere

The effects of GWs on the general circulation of the atmosphere above the turbopause in the thermosphere has been studied to a much lesser extent due to a combination of reasons. First, most middle atmosphere models have not extended into the upper atmosphere and upper atmosphere models were limited to simulating the thermosphere-ionosphere region. Additionally, previous GW parameterizations designed for the middle atmosphere were not developed for the whole atmosphere treatment of GWs. Explicit representation of small-scale GW effects in general circulation models extending from the lower atmosphere to the thermosphere-ionosphere requires extremely high model resolutions or appropriate GW parameterizations in order to represent the unresolved wave-induced effects (a_{gw}) on wind tendency seen in Eq. (5.28). A physically realistic representation of a_{gw} is required in order to realistically model Earth's atmosphere and other planetary atmospheres. Yiğit et al. (2009) have performed the first general circulation modeling study of the dynamical effects of lower atmospheric small-scale GWs on the general circulation of the thermosphere above the turbopause using the CMAT2 GCM coupled with the extended spectral nonlinear GW paramaterization of Yiğit et al. (2008). The use of the extended scheme have provided new capabilities of self-consistent propagation of GWs in the upper thermosphere and evaluation of direct wave effects. Figure 5.8 demonstrates their results for the dynamical effects of GWs on the neutral atmosphere, that is, the GW drag (upper panels), for June solstice conditions. The lower panels show the ion drag. The left column is the cut-off simulation while the right column is described as the extended simulation. These control simulations mimic the effect of "allowing" GW propagation into the thermosphere. Appreciable GW drag occurs above the turbopause as well and its magnitude is comparable to the GW effects in the middle atmosphere. GW drag in the thermosphere is strong, in particular, at middle- and high-latitudes in both hemispheres. Comparison with ion-neutral coupling effects (i.e., ion drag) suggests that the simulated GW drag competes with the ion drag up to F-region altitudes and the ion drag is greatly modulated by GW effects.

GW effects in the atmosphere are not limited to dynamical processes. Waves can heat and/or cool the fluid they propagate in. Figure 5.9 presents the mean zonal mean GW thermal effects resulting from their propagation and dissipation in the thermosphere at June solstice conditions. The direct GW heating (panel a) and the resultant total GW heating/cooling (panel b) are compared with important thermospheric thermal processes, the Joule heating (panel c) and molecular thermal conduction (panel d). The magnitude of GW thermal effects are appreciable and can even exceed Joule heating in the lower thermosphere.

A very high-resolution GCM that extends from the surface upward can to a large extent self-consistently include GW processes. Figure 5.10 shows the zonal mean zonal GW drag simulated at a very high-resolution of $1.1° \times 1.1°$ as presented in the work by Miyoshi et al. (2014) for the same solstice conditions assumed in Yiğit et al. (2009)'s GCM simulations. Intercomparison of high-resolution GW simulations and the parameterized GW simulations demonstrate a good agreement, suggesting that

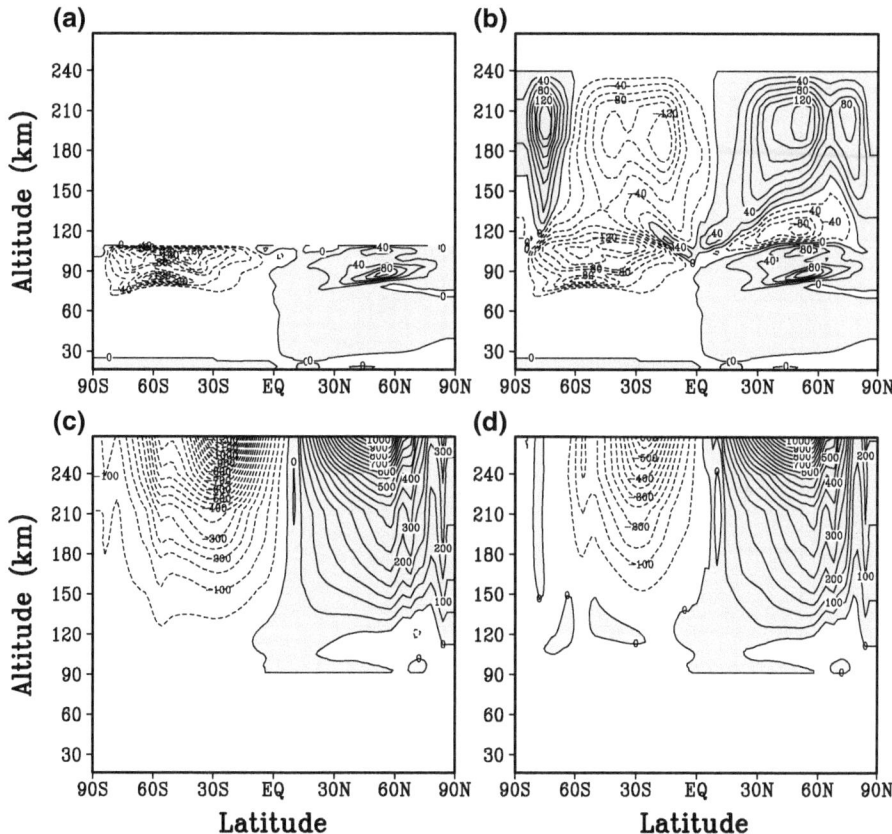

Fig. 5.8 Simulated gravity wave dynamical effects (*upper panels*) and ion drag (*lower panels*) with the CMAT2-GCM and the extended GW parameterization of Yiğit et al. (2009). Adopted from Yiğit et al. (2009, Fig. 3). **a** a_x^{GW} (EXP1). **b** a_x^{GW} (EXP2). **c** a_x^{ION} (EXP1). **d** a_x^{ION} (EXP2)

thermospheric effects of small-scale GWs of lower atmospheric origin can be adequately represented by GW parameterizations, specifically, by the extended nonlinear spectral parameterization of Yiğit et al. (2008).

In planetary atmospheres, GCMs are powerful tools in understanding coupling processes. On Mars, contemporary general circulation modeling efforts have already been performed in 1990s (e.g., Bougher et al. 1990). Recently, the extended scheme has been successfully used in the general circulation modeling of GW effects in the Martian atmosphere as well (Medvedev et al. 2011b; Medvedev and Yiğit 2012; Medvedev et al. 2013, 2015). These modeling efforts overall suggest appreciable GW effects on the general circulation of the Martian upper atmosphere. Figure 5.11 shows the simulation results with the Max Planck Institute Martian General Circulation Model of Medvedev and Yiğit (2012). They have demonstrated for the first time that accounting simultaneously for the thermal and dynamical effects of lower

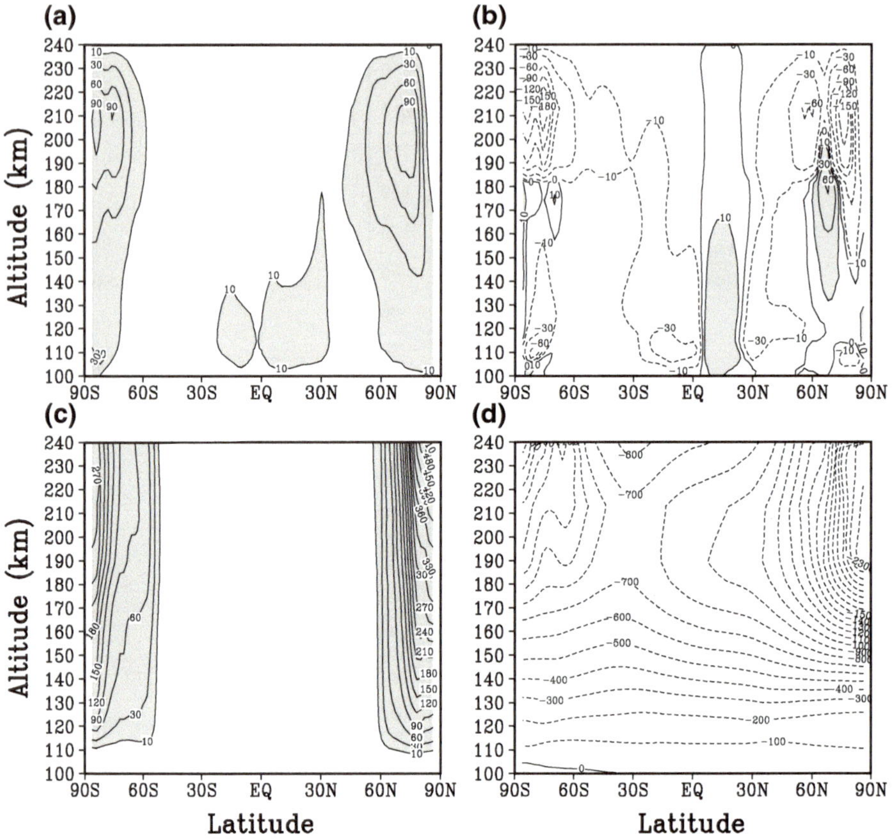

(a) **(b)** **(c)** **(d)**

Altitude (km) — plots (a), (b), (c), (d) with latitude axis 90S 60S 30S EQ 30N 60N 90N

Fig. 5.9 Simulated gravity wave thermal effects and comparison with Joule heating and thermal conduction. Adopted from Yiğit and Medvedev (2009, Fig. 2). **a** Irrevesible GW heating (E_1). **b** GW heating/cooling ($E_1 + Q_1$) **c** Joule heating. **d** Thermal cond.

Fig. 5.10 Simulation of gravity wave drag at high resolution with the Kyushu GAIA GCM without using GW parameterization. (Fig. 3, Miyoshi et al. 2014)

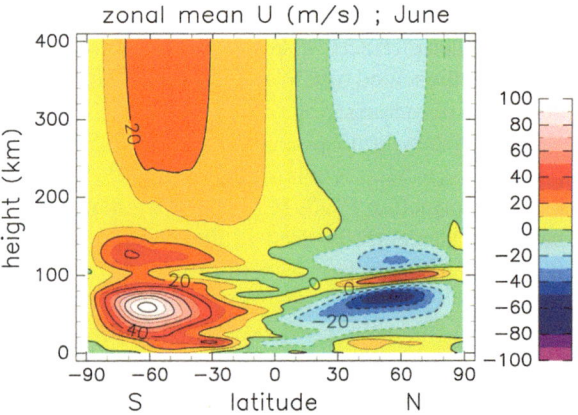

zonal mean U (m/s) ; June

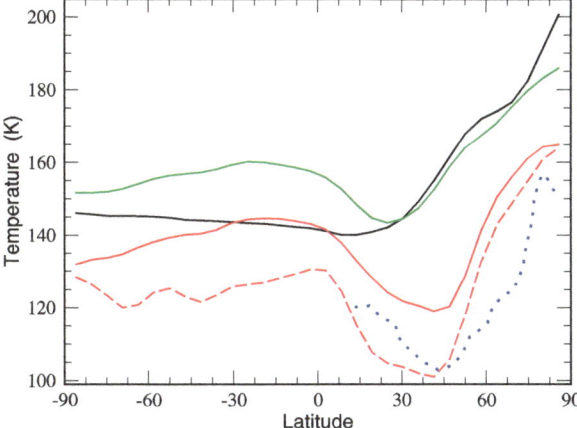

Fig. 5.11 Zonal mean neutral temperature at ~120 km simulated by the Max Planck Institute Mars General Circulation Model (MGCM) for various confugrations for the gravity waves. Run without GWs (*black*), with including only GW dynamical effects (*green*), and including both dynamical and thermal effects (*red*) are shown. The *blue dotted line* shows the inferred night-time neutral temperature from ODY aerobraking measurements published by Bougher et al. (2006). The *red dashed line* compares the associated modeled night-time temperature with bith GW dynamical and thermal effects accounted for (Fig. 2, Medvedev and Yiğit 2012)

atmospheric GWs on the Martian upper atmosphere enables the model to reproduce the night-time temperature profile inferred from ODY aerobraking measurements presented in the work by Bougher et al. (2006). Previous Martian modeling studies had been to a large extent unsuccessful reproducing the lower thermospheric temperatures on Mars.

References

Alexander MJ, Dunkerton TJ (1999) A spectral parameterization of mean-flow forcing due to breaking gravity waves. J Atmos Sci 56:4167–4182

Becker E (2011) Dynamical control of the middle atmosphere. Space Sci Rev 168:283–314. doi:10.1007/s11214-011-9841-5

Beres JH, Alexander MJ, Holton JR (2004) A method of specifying the gravity wave spectrum above convection based on latent heating properties and background wind. J Atmos Sci 61(3):324–337. doi:10.1175/1520-0469(2004)061

Bougher SW, Roble RG, Ridley EC, Dickinson RE (1990) The Mars thermosphere 2. general circulation with coupled dynamics and composition. J Geophys Res 95(B9):14,811–14,827

Bougher SW, Bell JM, Murphy JR, López-Valverde MA, Withers PG (2006) Polar warming in the Mars thermosphere: seasonal variations owing to changing insolation and dust distributions. Geophys Res Lett 33:L02203. doi:10.1029/2005GL024059

Chun HY, Baik JJ (2002) An updated parameterization of convectively forced gravity wave drag for use in large-scale models. J Geophys Res 59(5):1006–1017. doi:10.1175/1520-0469(2002)059

Curry MJ, Murty RC (1973) Thunderstorm-generated gravity waves. J Atmos Sci 31:1402–1408

Dewan EM, Good RE (1986) Saturation and the universal spectrum of vertical profiles of horizontal scalar winds in atmosphere. J Geophys Res 91:2742–2748

Djuth FT, Sulzer MP, Gonzales SA, Mathews JD, Elder JH, Walterscheid RL (2004) A continuum of gravity waves in the Arecibo thermosphere? J Geophys Res 31:L16801. doi:10.1029/2003GL019376

Djuth FT, Zhang LD, Livneh DJ, Seker I, Smith SM, Sulzer MP, Mathews JD, Walterscheid RL (2010) Arecibo's thermospheric gravity waves and the case for an ocean source. J Geophys Res 115:A08305. doi:10.1029/2009JA014799

Ern M, Preusse P (2012) Gravity wave momentum flux spectra observed from satellite in the summertime subtropics: implications for global modeling. Geophys Res Lett 39:L15810. doi:10.1029/2012GL052659

Forbes JM (1984) Middle atmosphere tides. J Atmos Terr Phys 46(11):1049–1067

Fritts DC (1984) Gravity wave saturation in the middle atmosphere: a review of theory and observations. Rev Geophys Space Phys 22:275–308

Fritts DC, Alexander MJ (2003) Gravity wave dynamics and effects in the middle atmosphere. Rev Geophys 41(1):1003. doi:10.1029/2001RG000106

Fritts DC, Wang L, Tolson RH (2006) Mean and gravity wave structures and variability in the Mars upper atmosphere inferred from Mars global surveyor and Mars odyssey aerobraking densities. J Geophys Res 111:A12304. doi:10.1029/2006JA011897

Gall RL, Williams RT, Clark TL (1988) Gravity waves generated during frontogenesis. J Atmos Sci 45:2204–2019

Garcia RR, Solomon S (1985) The effect of breaking gravity waves on the dynamics and chemical composition of the mesosphere and lower thermosphere. J Geophys Res 90:3850–3868 (implementation of Lindzen's parameterization into a two-dimensional dynamical model to study the effects of GWs in the MLT)

Garcia RR, Marsh DR, Kinnison DE, Boville BA, Sassi F (2007) Simulations of secular trends in the middle atmosphere. J Geophys Res 112:D09301. doi:10.1029/2006JD007485

Gille J, Barnett J, Arter P, Barker M, Bernath P, Boone C, Cavanaugh C, Chow J, Coffey M, Craft J, Craig C, Dials M, Dean V, Eden T, Edwards DP, Francis G, Halvorson C, Harvey L, Hepplewhite C, Khosravi R, Kinnison D, Krinsky C, Lambert A, Lee H, Lyjak L, Loh J, Mankin W, Massie S, McInerney J, Moorhouse J, Nardi B, Packman D, Randall C, Reburn J, Rudolf W, Schwartz M, Serafin J, Stone K, Torpy B, Walker K, Waterfall A, Watkins R, Whitney J, Woodard D, Young G (2008) High resolution dynamics limb sounder: experiment overview, recovery, and validation of initial temperature data. J Geophys Res: Atmos 113(D16). doi:10.1029/2007JD008824, http://dx.doi.org/10.1029/2007JD008824

Gossard EE, Hooke WH (1975) Waves in the atmosphere: atmospheric infrasound and gravity waves. Elsevier, Amsterdam

Hagan ME, Forbes JM (2002) Migrating and nonmigrating diurnal tides in the middle and upper atmosphere excited by tropospheric latent heat release. J Geophys Res 107(D24):4754. doi:10.1029/2001JD001236

Hagan ME, Forbes JM (2003) Migrating and nonmigrating semidiurnal tides in the middle and upper atmosphere excited by tropospheric latent heat release. J Geophys Res 108(A2):1062. doi:10.1029/2002JA009466

Hertzog A, Boccara G, Vincent RA, Vial F, Cocquerez P (2008) Estimation of gravity wave momentum flux and phase speeds from quasi-Lagrangian stratospheric balloon flights. Part II: Results from the Vorcore campaign in Antarctica. J Atmos Sci 65:3056–3070

Hickey MP, Cole KD (1988) A numerical model for gravity wave dissipation in the thermosphere. J Atmos Terr Phys 50:689–697

Hines CO (1960) Internal gravity waves at ionospheric heights. Can J Phys 38:1441–1481

Hines CO (1991) The saturation of gravity waves in the middle atmosphere. Part1: Critique of linear-instability theory. J Atmos Sci 48:1348–1359

Hines CO (1997a) Doppler spread parameterization of gravity wave momentum deposition in the middle: 1. Basic formulation. J Atmos Sol-Terr Phys 59:371–386

Hines CO (1997b) Doppler spread parameterization of gravity wave momentum deposition in the middle: 2. Broad and quasi monochromotic spectra and implemention. JASTP 59:387–400

Hines CO, Hooke WH (1970) Discussion of ionization effects on the propagation of acoustic-gravity waves in the ionosphere. J Geophys Res 75:2563–2568

Hoffmann L, Xue X, Alexander MJ (2013) A global view of stratospheric gravity wave hotspots located with atmospheric infrared sounder observations. J Geophys Res 118. doi:10.1029/2012JD018658

Holton JR (1982) The role of gravity wave induced drag and diffusion in the momentum budget of the mesosphere. J Atmos Sci 39:791–799

Holton JR (1983) The influence of gravity wave breaking on the general circulation of the middle atmosphere. J Atmos Sci 40:2497–2507

Holton JR, Wehrbein WM (1980) A numerical model of the zonal mean circulation of the middle atmosphere. Pure Appl Geophys 118:284–306

Hunt BG (1986) The impact of gravity wave drag and diurnal variability on the general circulation of the middle atmosphere. J Meteor Soc Jpn 64:1–16

Kim YJ, Eckermann SE, Chun HY (2003) An overview of the past, present and future of gravity-wave drag parametrization for numerical climate and weather prediction models. Atmos Ocean 41:65–98

Klostermeyer J (1972) Influence of viscosity, thermal conduction, and ion drag on the propagation of atmospheric gravity waves in the thermosphere. Z Geophysik 38:881–890

Leovy C (1964) Simple models of thermally driven mesospheric circulation. J Atmos Sci 21:327–341

Lindzen RS (1981) Turbulence and stress owing to gravity waves and tidal breakdown. J Geophys Res 86:9707–9714

Livneh DJ, Seker I, Djuth FT, Mathews JD (2007) Continuous quasiperiodic thermospheric waves over Arecibo. J Geophys Res 112:A07313. doi:10.1029/2006JA012225

Lott F, Miller MJ (1997) A new sub-grid orographic drag parameterization: its formulation and testing. J Geophys Res 102:26,053–26,076

Manzini E, McFarlane NA (1998) The effect of varying the source spectrum of a gravity wave parameterization in a middle atmosphere general circulation model. J Geophys Res 103:31,523–31,539

Matsuno T (1982) A quasi one-dimensional model of the middle atmosphere circulation interacting with internal gravity waves. J Meteor Soc Jpn 60:215–226

McFarlane NA (1987) The effect of orographically excited gravity wave drag on the general circulation of the lower stratosphere and troposphere. J Atmos Sci 44:1775–1800

Medvedev AS, Gavrilov NM (1995) The nonlinear mechanism of gravity wave generation by meteorological motions in the atmosphere. J Atmos Terr Phys 57:1,221–1,231

Medvedev AS, Klaassen GP (1995) Vertical evolution of gravity wave spectra and the parameterization of associated wave drag. J Geophys Res 100:25,841–25,853

Medvedev AS, Yiğit E (2012) Thermal effects of internal gravity waves in the Martian upper atmosphere. Geophys Res Lett 39:L05201. doi:10.1029/2012GL050852

Medvedev AS, Yiğit E, Hartogh P (2011a) Estimates of gravity wave drag on Mars: indication of a possible lower thermosphere wind reversal. Icarus 211:909–912. doi:10.1016/j.icarus.2010.10.013

Medvedev AS, Yiğit E, Hartogh P, Becker E (2011b) Influence of gravity waves on the Martian atmosphere: general circulation modeling. J Geophys Res 116:E10004. doi:10.1029/2011JE003848

Medvedev AS, Yiğit E, Kuroda T, Hartogh P (2013) General circulation modeling of the Martian upper atmosphere during global dust storms. J Geophys Res Planets 118:1–13. doi:10.1002/jgre.20163

Medvedev AS, González-Galindo F, Yiğit E, Feofilov AG, Forget F, Hartogh P (2015) Cooling of the martian thermosphere by CO_2 radiation and gravity waves: an intercomparison study with two general circulation models. J Geophys Res Planets 120. doi:10.1002/2015JE004802, http://dx.doi.org/10.1002/2015JE004802

Miyoshi Y, Fujiwara H, Jin H, Shinagawa H (2014) A global view of gravity waves in the thermosphere simulated by a general circulation model. J Geophys Res Space Phys 119:5807–5820. doi:10.1002/2014JA019848

Munro (1950) Traveling disturbances in the ionosphere. Proc R Soc London A 202(1069):208–223

Nappo CJ (2002) An introduction to atmospheric gravity waves. International geophysics series, vol 85. Academic Press, London

Palmer TN, Shutts GJ, Swinbank R (1986) Alleviation of a systematic westerly bias in general circulation and numerical weather prediction models through an orographic gravity wave drag parameterization. Q J R Meteorol Soc 112:1001–1039

Park J, Lühr H, Lee C, Kim YH, Jee G, Kim JH (2014) A climatology of medium-scale gravity wave activity in the midlatitude/low-latitude daytime upper thermosphere as observed by CHAMP. J Geophys Res Space Phys 119. doi:10.1002/2013JA019705

Pitteway MLV, Hines CO (1963) The viscous damping of atmospheric gravity waves. Can J Phys 41:1935–1948

Schirber SEM, Alexander MJ (2014) A convection-based gravity wave parameterization in a general circulation model: implementation and improvements on the QBO. J Adv Model Earth Syst 6:264–279. doi:10.1002/2013MS000286

Schoeberl MR, Strobel DF (1978) The zonally averaged circulation of the middle atmosphere. J Atmos Sci 35:577–591

Shume EB, Rodrigues FS, Mannucci AJ, de Paula ER (2014) Modulation of equatorial electrojet irregularities by atmospheric gravity waves. J Geophys Res Space Phys 119. doi:10.1002/2013JA019300

Smith AK (2012) Global dynamics of the MLT. Surv Geophys 33. doi:10.1007/s10712-012-9196-9

Song IS, Chun HY, Lane TP (2003) Generation mechanisms of convectively forced internal gravity waves and their propagation to the stratosphere. J Atmos Sci 60:1960–1980

Song IS, Chun HY, Garcia RR, Boville BA (2007) Momentum flux spectrum of convectively forced internal gravity waves and its application to gravity wave drag parameterization. part ii: Impacts in a GCM (WACCM). J Atmos Sci 34:2286–2308

Vadas S, Liu H (2009) Generation of large-scale gravity waves and neutral winds in the thermosphere from the dissipation of convectively generated gravity waves. J Geophys Res 114:A10310. doi:10.1029/2009JA014108

Vadas SL, Crowley G (2010) Sources of the traveling ionospheric disturbances observed by the ionospheric tiddbit sounder near Wallops Island on 30 October 2007. J Geophys Res 115:A07324. doi:10.1029/2009JA015053

Vadas SL, Fritts DC (2005) Thermospheric responses to gravity waves: influences of increasing viscosity and thermal diffusivity. J Geophys Res 110:D15103. doi:10.1029/2004JD005574

Vadas SL, Fritts DC, Alexander MJ (2003) Mechanism for the generation of secondary waves in wave breaking regions. J Atmos Sci 60:194–214

VanZandt TE (1982) A universal spectrum of buoyancy waves in the atmosphere. Geophys Res Lett 9:575–578

Warner CD, McIntyre ME (2001) An ultrasimple spectral parameterization for nonorographic gravity waves. J Atmos Sci 58:1837–1857

Weinstock J (1976) Nonlinear theory of acoustic-gravity waves 1. Saturation and enhanced diffusion. J Geophys Res 81:633–652

Weinstock J (1982) Nonlinear theory of gravity waves: momentum deposition, generalized rayleigh friction, and diffusion. J Atmos Sci 39:1,698–1,710

Witt G (1962) Height, structure and displacements of noctilucent clouds. Tellus 1–18:1–2

Yiğit E, Medvedev AS (2009) Heating and cooling of the thermosphere by internal gravity waves. Geophys Res Lett 36:L14807. doi:10.1029/2009GL038507

Yiğit E, Medvedev AS (2010) Internal gravity waves in the thermosphere during low and high solar activity: simulation study. J Geophys Res 115:A00G02. doi:10.1029/2009JA015106

Yiğit E, Medvedev AS (2015) Internal wave coupling processes in Earth's atmosphere. Adv Space Res 55:983–1003. doi:10.1016/j.asr.2014.11.020, http://www.sciencedirect.com/science/article/pii/S0273117714007236

Yiğit E, Aylward AD, Medvedev AS (2008) Parameterization of the effects of vertically propagating gravity waves for thermosphere general circulation models: sensitivity study. J Geophys Res 113:D19106. doi:10.1029/2008JD010135

Yiğit E, Medvedev AS, Aylward AD, Hartogh P, Harris MJ (2009) Modeling the effects of gravity wave momentum deposition on the general circulation above the turbopause. J Geophys Res 114:D07101. doi:10.1029/2008JD011132

Chapter 6
Atmospheric Circulation and Dynamical Processes

Basic Principles and a Case Study

The motion of the surface of water resembles the behaviour of hair, which has two motions, of which one depends on the weight of the strands, the other on the line of its revolving; thus water makes revolving eddies, one part of which depends upon the impetus of the principle current, and the other depends on the incident and reflected motions.

—Leonardo da Vinci (16th century A.D.)

Abstract Fundamentals of eddies and atmospheric circulation are briefly presented. Atmospheric circulation patterns, the large-scale general circulation and diagnostic techniques for characterizing atmospheric flow, such as Transformed Eulerian framework, are discussed. Basic features of general circulation models are introduced and their application in studying atmospheric circulation and coupling processes are illustrated in the context of sudden stratospheric warmings and atmospheric gravity wave propagation and dissipation.

Keywords Circulation · Angular momentum · General circulation · Radiative equilibrium · Atmospheric circulation · Eddy motion · Transformed eulerian mean · Sudden stratospheric warming · Gravity waves

6.1 Eddy and Vortex

Atmospheric eddies play a great role in shaping the large-scale atmospheric circulation and the associated dynamical processes, despite their relatively small scales. The phenomenon of small-scale structures driving/controlling large-scale processes is, admittedly speaking, relatively unintuitive. What are eddies and how are their roles quantified in studies of atmospheric dynamics? Or in other words, what is the role of small-scale processes in the large-scale flow? Small-scale internal waves are an excellent illustration of how the large-scale atmospheric dynamics is driven by small-scale processes. Dynamical impact of these waves on the atmospheric flow are summarized in a number of reviews (e.g., Fritts 1984; Fritts and Alexander 2003; Laštovička

© The Author(s) 2015
E. Yiğit, *Atmospheric and Space Sciences: Neutral Atmospheres*,
SpringerBriefs in Earth Sciences, DOI 10.1007/978-3-319-21581-5_6

2006; Becker 2011; Smith 2012; Yiğit and Medvedev 2015). The global effects of small-scale vertical disturbances are presented in the work by Smith (2000). Also, small-scale structures of lower atmospheric origin propagating upward have great implications for the spatiotemporal variations of the middle and upper atmosphere (Mengel et al. 1995; Fritts et al. 2006; Yiğit and Ridley 2011). These questions will be a central part of this chapter.

Let us first discuss eddies in the context of a vortex. An eddy can be illustrated by a small vortex structure. The theoretical concepts underlying vortices have already been developed in the 19th century primarily by H. Helmholtz and Kelvin. The first investigation of a vortex motion has been done in the work by Helmholtz (1858) by introducing *vortex lines* in the fluid. Helmholtz defines them in the original German publication as "Wirbellinien". A "Wirbel" is a vortex and a "Linie" is a line. A vortex line is a theoretical line that shows everywhere the axis of rotation of the fluid. *Vortex tubes* describe the hypothetical volumes that are formed by imagining closed curves through which the vortex lines go. Lamb (1932) describes a vortex simply as a rotational motion and defines it as the fluid that a vortex tube contains.

An eddy, sometimes called a turbulent eddy, is associated with a vortex motion of the velocity field **u**. Vorticity ω is defined as a rotational velocity field

$$\omega = \nabla \times \mathbf{u}. \tag{6.1}$$

Consider a two-dimensional flow, for example, a horizontal flow in the meridional and zonal direction, then the vorticity is given by

$$\omega = \left(0,\ 0,\ \frac{\partial v}{\partial x} - \frac{\partial u}{\partial y}\right). \tag{6.2}$$

Figure 6.1 illustrates the vorticity of a random horizontal flow of the structure

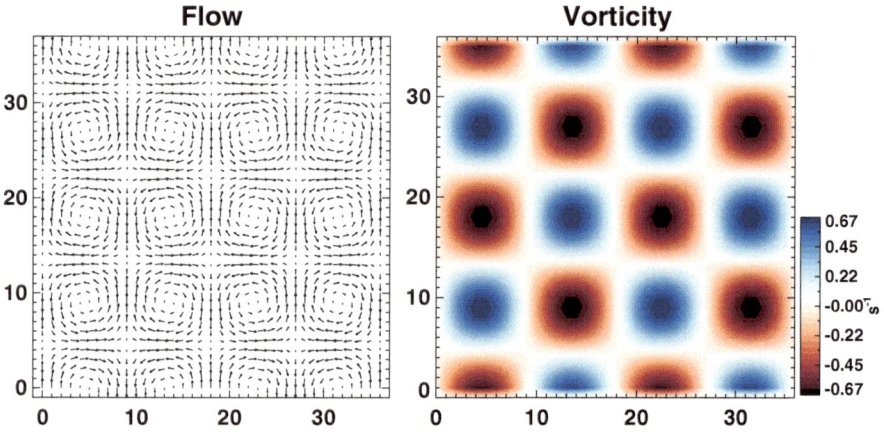

Fig. 6.1 Vorticity field of a two-dimensional flow

$$\mathbf{u} = A \left(\sin kx \, \sin ly, \cos kx \, \cos ly, 0 \right) \tag{6.3}$$

with amplitude of unity and with wave number two in both directions. Left panel shows the velocity vector field and the right panel shows the associated distribution of the vorticity field in the z-direction, i.e., ω_z. It can be seen that in places of rotational motion the magnitude of vorticity is large, while in cases of a straight flow, there is no vorticity. Vorticity has indeed a sign depending on whether it is a clockwise or anticlockwise rotation. In the presence of higher-wavenumber structures in a velocity field, smaller scale vortices can occur. One of the aspects of complexity in atmospheric dynamics is that the velocity field exhibits a continuous eddy motion that rapidly evolves in space and time.

Eddies indeed have practical implications for the description of atmospheric motion. In the past, in order to simplify the analysis, primarily zonally averaged flows have been studied (e.g., Andrews and McIntyre 1976; Schoeberl and Strobel 1978; Holton and Wehrbein 1980). However, in a realistic atmosphere, the time-averaged circulation is highly dependent on the longitude due to longitudinally asymmetric forcing by orography and/or land-sea heating contrasts. To better understand the dynamics of the atmosphere, it is necessary to isolate the processes that maintain the zonal mean flow, $\bar{u}$. Linear wave studies allow the decomposition of the actual flow into a zonal mean flow $\bar{u}$ and a spatially (e.g., longitudinally) dependent "eddy" components u' as

$$u = \bar{u} + u'. \tag{6.4}$$

This decomposition is also called *Reynold's decomposition*. Sometimes other notations, such as, $u = u_0 + u_1$ and $u = <u> + u_1$, are used as well.

It is of great interest to determine the influence of eddies on the structure of the atmospheric flow. Atmospheric waves, e.g., acoustic waves, gravity waves, Rossby waves, and tides, can produce eddies of varying scales in the atmosphere when they become dynamically or convectively unstable.

6.2 Radiative Equilibrium

We have to remind ourselves that Earth's atmosphere can be viewed as a stably stratified fluid in which radiative, electromagnetic, dynamical and thermal processes occur simultaneously. The interplay of these processes offers a substantial amount of complexity in the system. One of the simplest representations of a planetary atmosphere can be done by assuming radiative equilibrium conditions. The incoming solar radiation would heat the atmosphere and the outgoing infrared radiation would be the only cooling mechanism. In thermal equilibrium, the rate of radiation absorption would be equal to the rate at which the energy is emitted. In other words, no other physical mechanism would be contributing to the transport of heat. In fact, Earth is neither an ideal black body nor an ideal reflector. Its net energy balance departs radically from a radiative equilibrium because of a dense atmosphere. For example, owing to the presence of CO_2 and water vapor in the atmosphere, Greenhouse effect can take place, leading to much warmer surface temperatures than without CO_2.

6.3 Concept of Circulation

Characterization of atmospheric circulation is a central aspect of dynamical meteorology. In order to appreciate the concept of circulation, let's first discuss angular momentum given by

$$\mathbf{L} = \mathbf{r} \times \mathbf{p},\tag{6.5}$$

where $\mathbf{r}$ and $\mathbf{p} = m\mathbf{u}$ are the position and linear momentum of a given object. The angular momentum is perpendicular to the plane formed by the position and linear momentum vectors. The conservation of angular momentum is a powerful constraint on the behavior of rotational motion. The physical significance of the angular momentum lies in its magnitude and direction, not in its specific location and it depends on the choice of the reference point (Kleppner and Kolenkow 1973). Similarly, the notions of circulation and vorticity provide a measure of rotation in a fluid, where circulation is a macroscopic property, while vorticity in principle is of microscopic nature.

Circulation involves motion and a closed path. It can be formally represented by the Circulation theorem, which states that the circulation about a closed contour in a fluid is the line integral of the velocity vector evaluated tangent to the contours

$$C = \oint \mathbf{u} \cdot d\mathbf{l}.\tag{6.6}$$

By convention $C > 0$ is counterclockwise integration and $C < 0$ is clockwise integration. For a barotropic fluid, i.e., $\rho = \rho(p)$, the absolute circulation is conserved following the motion. The advantage of working with C is that no reference is needed to an axis of rotation.

There is a variety of atmospheric circulation types. At equinox and solstice, differential solar heating produces a meridional temperature gradient, which sets up different global circulation patterns. The *diabatic circulation* is the dynamical circulation that balances this differential heating (Andrews et al. 1987). The *zonal mean circulation* is zonally averaged (i.e., around a longitude circle) flow with the zonal mean zonal ($\bar{u}$), meridional ($\bar{v}$) and vertical ($\bar{w}$) components. The *residual mean circulation* $(0, \bar{v}^*, \bar{w}^*)$ is a diagnostic mean circulation of the zonal mean flow as a function of altitude and latitude, incorporating the effects of eddies on the flow (see Sect. 6.8). The *general circulation* of the atmosphere is a rather large-scale flow of the atmosphere, which we will discuss in detail in the next section.

6.4 General Circulation

The *general* (or *global*) *circulation* of the atmosphere can be understood as total global-scale variation of atmospheric flow. This expression is typically used to describe the large-scale circulation of the neutral fluid. A variety of dynamical, thermal, and electrodynamical processes influence the general circulation. In order to precisely determine the circulation, "all" physical processes acting on the energy, momentum, and composition balance of a given air parcel ought to be determined. As for the neutrals, all three conservation laws (Sects. 3.5–3.7) have to be satisfied simultaneously, the processes of determining atmospheric circulation involves solving a set of coupled partial differential equations in time and space. A better estimate of the tendencies of velocity, temperature, and mass flow provides a more precise description of atmospheric general circulation. Such a complex task requires the application of general circulation models, which were mentioned briefly in Sect. 5.10.1.1 in the context of quantifying gravity wave effects in planetary atmospheres.

6.5 General Circulation Modeling

Modeling physical processes involves numerical computing. The art of computing is in fact as old as mathematics and logic. Computing aids have been developed already in the ancient times. In 17th century Pascal and Leibniz designed physical devices that could perform arithmetic calculations mechanically. The modern concepts behind complex atmospheric modeling can be traced back to early 1900s. Bjerknes (1904) discussed in his seminal publication the basics of methods of solving mathematical equations of atmospheric and oceanic flows. During the first World War, L.F. Richardson has used basic equations of atmospheric physics to predict weather systems, using a mechanical calculator. He also showed how differential equations could be written as a set of algebraic difference equations for values of the tendencies of atmospheric field variables at finite number of grid points in space (Richardson 1922). Yet, advanced numerical calculations required electronic computers, whose development started during the second World War in 1940s. Ballistics tables that aid in positioning artillery pieces and in making bombing raids effective have become strategically very significant during the World War II. Such complex tables took a long time to calculate, as a large number of variables had to be taken into account. Therefore, scientists created the world's first electronic, large-scale digital computer named ENIAC (Electronic Numerical Integrator and Calculator). Its sole job was to calculate trajectories under different conditions. This machine took a huge amount of space. John von Neumann has realized the weakness of having a huge machine to perform a single task. He designed the concept of a stored-program computer, in which a set of operations to be performed could be input. This concept gave rise to the first general purpose stored-program computer, the EDVAC (Electronic Discrete Variable Automatic Computer), forming the foundation of the modern computer.

Nowadays, rapid technological advancements, continuous and global observations motivate the development of sophisticated complex general circulation models (GCMs). Earlier models used to be two-dimensional, but, today, three-dimensionality is an established characteristics of all GCMs around the world.

The major components of GCMs are land, sea, atmosphere-ionosphere, and sea ice. It is possible to have coupled GCMs, such as, an atmosphere-ocean GCM. In this book, I am focusing on atmospheric GCMs. Models are based on the iterative numerical solutions of coupled governing partial differential equations of atmospheric dynamics in a specified horizontal and vertical grid. For a set of initial and boundary conditions tendencies for velocities, temperature, and composition are evaluated. The solution method involves various numerical techniques. As the governing equations are in the form of differential equations, they cannot be directly represented in models. These equations thus have to be discretized, for example, by replacing them by difference equations, as it is done in the finite difference technique. The difference equations can then be solved iteratively in order to calculate model tendencies. For example, the unsteady heat conduction equation with constant thermal diffusivity (2.54) can be disretized as

$$\frac{T_i^{n+1} - T_i^n}{\Delta t} \approx \frac{T_{i+1}^n - 2T_i^n + T_{i-1}^n}{(\Delta x)^2}, \tag{6.7}$$

where "n" and "i" are the time and space indices, and the individual variables represent

$$\begin{aligned}
T_i^n &= T(x, t) \\
T_i^{n+1} &= T(x, t + \Delta t) \\
T_{i+1}^n &= T(x + \Delta x, t) \\
T_{i-1}^n &= T(x - \Delta x, t) \\
\Delta x &= x_{i+1} - x_i \\
\Delta t &= t^{n+1} - t^n
\end{aligned} \tag{6.8}$$

In the above approximation, first-order forward difference approximation in time and second-order central difference scheme in space are used. The subscript and superscript notations in Eq. (6.8) are quiet useful because the variations of temperature can be represented more conveniently. Given n is the current time step, from (6.7) the value of temperature in the next time step $n + 1$ can be determined.

$$T_i^{n+1} \approx \frac{\Delta t}{(\Delta x)^2}(T_{i+1}^n - 2T_i^n + T_{i-1}^n) + T_i^n. \tag{6.9}$$

The associated grid is illustrated in Fig. 6.2 with respect to the grid point (i, n). The value of temperature at grid point $(i, n + 1)$ can be approximated explicitly in terms of the temperature value at $(i - 1, n)$, (i, n), and $(i + 1, n)$.

Fig. 6.2 An illustration of a time-space grid, where Δx and Δt denote the steps in space and time. The point (i, n) is the reference grid point with respect to which the other points are defined

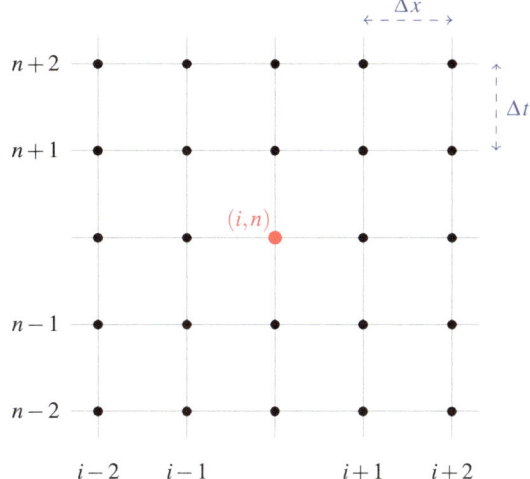

Once appropriate boundary conditions are specified, the numerical solutions can be successfully performed provided that numerical *stability, consistency*, and *convergence* are ensured. For example, Courant et al. (1928) have suggested that there exists a numerical constraint to have a stable numerical solution for finite difference equations. Their constrain known as the Courant-Friedrichs-Lewy criterion established a ratio of the spatial to the temporal resolution and the speed of the fastest propagating wave in the fluid system. If numerical instability occurs, then model solutions exhibit exponential growth with time and atmospheric processes cannot be realistically simulated.

Depending on the vertical extent of GCMs, there are lower atmosphere, middle atmosphere and upper atmosphere GCMs. The middle atmosphere GCMs typically extend from the surface up to the mesopause (e.g., Beagley et al. 1997, while upper atmosphere models include the thermosphere and/or the ionosphere only (e.g., Fuller-Rowell et al. 1996; Ridley et al. 2006; Gardner and Schunk 2011). The general tendency in the modeling community is to develop whole-atmosphere GCMs (e.g., Garcia et al. 2007; Jin et al. 2011; Liu et al. 2013; Miyoshi et al. 2014). Some of these models provide a self-consistent two-way coupling between the ionosphere and the atmosphere, providing an unprecedented global view of the atmosphere-ionosphere system from the ground to the exobase. Including more atmospheric regions and physics requires stronger computational resources. The use of physics-based parameterizations could be a valuable asset for whole-atmosphere models, as they provide a practical and computationally less expensive formulation of physical processes that are unresolved in models. Future GCMs are expected include even space weather effects in a self-consistent manner. This development theoretically requires a self-consistent magnetosphere coupled to the effects of Sun. Increasing amount of global observations can place better constrains on GCM simulations.

6.6 Circulation Patterns in the Atmosphere

Edmond Halley (17th cc) and George Hadley (18th cc) were two of the first scientists who worked on the circulation of the atmosphere. The Hadley circulation was characterized by a single cell pattern and zonal symmetry. According to Hadley, the solar heating at the equator causes air to rise, the risen air then moves poleward where it sinks again and travels back to the equator. This model involves vertical circulation in each hemisphere.

Following Hadley's hypothesis of single-cell atmospheric circulation, Ferrel suggested a three-cell pattern. His model was described by the following: (1) Air rises near equator and descends near 30°; (2) air is descending near 30° and rising at 60° (3) air rises again at 60° and descends near the poles, forming the polar cell. He also suggested that the Coriolis force would produce a deflection effect on the neutral wind flow. That is, air traveling from the equator to the pole gets deflected to the right, so no single-cell pattern would be possible. This concept could explain the counterclockwise rotation of cyclones (major low pressure weather systems) in the Northern Hemisphere and clockwise rotation of cyclones in the Southern Hemisphere via the Coriolis effects.

Both the Hadley and polar cells are direct because they are driven by heating and cooling patterns. By contrast the Ferrel cell is driven indirectly by the circulation system, in particular by the other two cells. There are significant variations from the basic form of these circulation patterns, because of the seasonal and land-sea contrasts. Seasonally, in the Northern Hemisphere summer, the three-cell patter shifts northward and southward during Southern Hemisphere summer. Another aspect that requires consideration is that the radiative heating distribution is seasonally dependent. Maximum heating occurs at the summer pole and maximum cooling at the winter pole. At equinox, maximum heating takes place at the equator and there is a resultant cooling at the poles. The so-called diabatic circulation is the meridional circulation that dynamically balances this differential heating. Therefore, overall, the atmosphere demonstrates great differences in terms of the changes in circulation pattern in different seasons.

6.7 Eulerian Mean

Dynamics of the middle atmosphere can be understood to a good extent by considering the interactions of eddies with the mean flow $\bar{u}$. Waves can significantly change the configuration of the mean flow and in turn, alteration in the mean flow modulates the propagation and dissipation characteristics of waves. Therefore, eddy-mean flow or wave-mean flow interactions are a two-way process that is largely nonlinear in nature. This nonlinearity is demonstrated mathematically by the structure of the governing equations of atmospheric dynamics and thermodynamics as presented in Sect. 3.9.

Complex numerical models, such as GCMs, can be used to solve the prognostic equations in three dimensions and time. However, for certain dynamical problems, it is convenient to have techniques that provide a simplified first-order diagnostics of how variations in the flow affect the totality of the flow. Although, the realistic atmospheric flow is three-dimensional, i.e., $\mathbf{u} = (u, v, w)$, we can simplify the treatment of dynamics in the lower and middle atmosphere by evaluating a zonally averaged flow, which is sometimes called the mean flow. It is denoted by an overbar " ‾ ". This averaging reduces the problem to two dimensions. In general, the time-dependent zonal mean of a field variable χ is given by

$$\bar{\chi}(z, \theta, t) = \frac{1}{2\pi} \int_0^{2\pi} \chi(z, \phi, \theta, t) \, d\phi, \tag{6.10}$$

where z is the vertical coordinate, ϕ and θ are longitude and latitude, respectively. This averaging is also called the *Eulerian mean*. Numerically, for N number of longitudes, the zonal mean is

$$\bar{\chi} = N^{-1} \sum_{i=1}^{N} \chi_i \tag{6.11}$$

The associated fluctuation χ' in χ is defined as a departure from $\bar{\chi}$, the zonal mean of χ

$$\chi'(z, \phi, \theta, t) \equiv \chi - \bar{\chi}. \tag{6.12}$$

Here, in order to illustrate the application of the Eulerian mean, consider the addition and multiplication of two variables χ and ψ. Addition yields

$$\begin{aligned} \overline{\chi + \psi} &= \overline{\bar{\chi} + \chi' + \bar{\psi} + \psi'} \\ &= \bar{\bar{\chi}} + \bar{\chi'} + \bar{\bar{\psi}} + \bar{\psi'} \\ &= \bar{\chi} + \bar{\psi}, \end{aligned} \tag{6.13}$$

where we have used the averaging properties $\bar{\bar{\chi}} = \bar{\chi}$ and $\bar{\chi'} = 0$, that is, the average of an average of a parameter is just the average of the parameter and the average of a fluctuation is zero. For multiplication we have

$$\begin{aligned} \overline{\chi \psi} &= \overline{(\bar{\chi} + \chi') \cdot (\bar{\psi} + \psi')} \\ &= \overline{\bar{\chi}\bar{\psi} + \bar{\chi}\psi' + \chi'\bar{\psi} + \chi'\psi'} \\ &= \bar{\chi}\bar{\psi} + \overline{\chi'\psi'}. \end{aligned} \tag{6.14}$$

The quantity of $\overline{\chi'\psi'}$ has wide-reaching physical significance besides its mathematical beauty. It is used to describe a possible correlation between different points in the flow. In reality, these fluctuation terms can be measurable quantities. For example, consider velocity fluctuations in a geophysical flow. In the turbulence theory, the velocity correlation function is $C_{ij} = \langle u_i'(\mathbf{r})\, u_j'(\mathbf{r}+d\mathbf{r})\rangle$, where the notation "$\langle\ \rangle$" denotes an ensemble average, that is, an average over many realizations, the subscript represent different velocity components. If there is no correlation between the velocity fluctuations u_i' and u_j', then $C_{ij} = 0$.

Separation of variables in a zonally averaged component and a disturbance has been found to be a satisfactory tool in the description of middle atmosphere dynamics (Andrews et al. 1987). Applying this decomposition to momentum, energy, and mass balance equations and then zonally averaging yield the primitive equations for the Eulerian mean flow, which form a closed set of equations for the temporal evolution of the zonal mean circulation (Andrews et al. 1987). The Eulerian mean flow equations include eddy forcing terms in the form of $\overline{\chi'\psi'}$ and the the mean meridional circulation consists of zonal mean vertical and meridional velocity components $(\bar{v}, \bar{w})$. This formalism is applicable to small amplitude disturbances in the mean flow. Large amplitude cases may require numerical simulations. A form of the dynamical equations for the Eulerian mean flow as presented in the book by Holton and Hakim (2012) is

$$\frac{\partial \bar{u}}{\partial t} - f_0 \bar{v} = -\frac{\partial (\overline{u'v'})}{\partial y} + \bar{X} \tag{6.15}$$

$$\frac{\partial \bar{T}}{\partial t} + \underbrace{N^2 \frac{H}{R} \bar{w}}_{\text{Adiabatic cooling}} = \underbrace{-\frac{\partial (\overline{v'T'})}{\partial y}}_{\substack{\text{Eddy heat} \\ \text{flux convergence}}} + \frac{\bar{J}}{c_p} \tag{6.16}$$

where the squared buoyancy frequency is

$$N^2 = \frac{R}{H}\left(\frac{\kappa T_0}{H} + \frac{dT_0}{dz}\right), \tag{6.17}$$

and $\overline{u'v'}$ and $\overline{v'T'}$ are eddy momentum and heat fluxes, respectively. For steady-state mean flow, the $(\bar{v}, \bar{w})$ circulation must just balance the eddy forcing plus diabatic heating so that the balances in Eqs. (6.15) and (6.16) remain valid. That is, Coriolis force approximately balances the divergence of eddy momentum fluxes and adiabatic cooling balances the diabatic heating plus convergence of eddy heat fluxes. Therefore, any changes in the zonal mean flow arise from small imbalances between the forcing terms and the mean meridional circulation.

6.8 Transformed Eulerian Mean (TEM)

Transformed Eulerian Mean (TEM) equations are an alternative analysis of the zonal mean circulation. Their application produces a clearer diagnosis of eddy forcing and provides a more direct view of transport processes in the meridional plane

(Andrews and McIntyre 1976). The problem of strong cancellation between adiabatic cooling and eddy heat flux convergence in (6.16) is alleviated. According to this theory, a residual mean circulation $(\bar{v}^*, \bar{w}^*)$, composed of the meridional components $\bar{v}^*$ and the vertical component $\bar{w}^*$, is defined as

$$\bar{v}^* = \bar{v} - \frac{1}{\rho_o} \frac{R}{H} \frac{\partial}{\partial z} \left(\frac{\rho_0 \overline{v'T'}}{N^2} \right) \tag{6.18}$$

$$\bar{w}^* = \bar{w} + \frac{R}{H} \frac{\partial}{\partial y} \left(\frac{\overline{v'T'}}{N^2} \right) \tag{6.19}$$

With the above substitutions, TEM equations can be obtained (Holton and Hakim 2012).

$$\frac{\partial \bar{u}}{\partial t} - f_0 \bar{v}^* = \overbrace{\frac{1}{\rho_0} \nabla \cdot \mathbf{F}}^{\text{Large-scale eddies}} + \overbrace{\tilde{\chi}}^{\text{Small-scale eddies}} \tag{6.20}$$

$$\frac{\partial \bar{T}}{\partial t} + N^2 \frac{H}{R} \bar{w}^* = \frac{\bar{J}}{c_p} \tag{6.21}$$

$$\frac{\partial \bar{v}^*}{\partial y} + \frac{1}{\rho_0} \frac{\partial (\rho_0 \bar{w}^*)}{\partial z} = 0, \tag{6.22}$$

where f_0 is the midlatitude Coriolis parameter, ρ_0 is the basic density, and $\mathbf{F}$ is the Eliassen-Palm Flux with the components

$$\mathbf{F} = (F_y, F_z) \tag{6.23}$$

$$F_y = \rho_0 f_0 R \frac{\overline{v'T'}}{N^2 H} \tag{6.24}$$

$$F_z = -\rho_0 \overline{u'v'}, \tag{6.25}$$

$\nabla \cdot \mathbf{F}$ is the Eliassen-Palm (EP) flux divergence (eddy divergence) and $\bar{\chi}$ incorporate the effects of small-scale (unresolved) waves. The total zonal mean force $\bar{G}$ acting on the mean flow due to both the large- and small-scale eddies is

$$\bar{G} = \frac{1}{\rho_0} \nabla \cdot \mathbf{F} + \bar{\chi} \tag{6.26}$$

The eddy heat flux $\overline{v'T'}$ and eddy momentum flux $\overline{u'v'}$ act together to drive the changes in the zonal mean circulation. These equations reveal that the fundamental role of eddies is to exert a zonal (body) force on the atmospheric circulation, where the physical meaning of the EP-flux divergence is a zonal body force per unit mass exerted by quasi-geostrophic eddies. Unlike the conventional Eulerian mean formalism, the residual mean vertical motion is for time-averaged conditions. The rate of diabatic heating represents approximately the diabatic circulation in the meridional plane. The time-averaged residual mean meridional circulation represents the mean motions of air parcels, which is an approximation to the mean advective transport of trace species. Mathematically, TEM equations can be formulated in terms of other vertical coordinates as well. In evaluating the EP flux and its divergence, one should choose a vertical coordinate system that is consistent with the vertical vertical coordinates in which atmospheric parameters are obtained. For example, if the field variables are as a function of spherical log-pressure coordinates, then the TEM equations should be expressed in this coordinate system in order to perform the EP flux diagnostics.

A number of authors have used the EP-flux diagnostics in order to study the wave-mean flow interactions, in particular, in the lower and middle atmosphere (Andrews and McIntyre 1976; Edmon et al. 1980; Dunkerton et al. 1981; Palmer 1981; Liebermann 1999). The EP-flux formalism is consistent with the nonacceleration theorem first introduced by Charney and Drazin (1961) can be stated as

$$\frac{\partial \bar{u}}{\partial t} = 0 \quad \Longleftrightarrow \quad \bar{v}^* = \bar{w}^* = 0. \tag{6.27}$$

Under the conditions of steady linear disturbances, the flow is conservative. A generalized form of the Eliassen-Palm Theorem is given by

$$\frac{\partial A}{\partial t} + \mathbf{\nabla} \cdot F = D + \mathcal{N}, \tag{6.28}$$

where the first term on the left-hand side represents the wave activity transience, D is the frictional/diabatic effects, and $\mathcal{N}$ includes nonlinear processes. The generalized EP-flux theorem states how the large-scale eddy effects depend on wave transience and nonconservative wave effects.

6.9 Sudden Stratospheric Warming

In the lower atmosphere, the troposphere is a source of various waves (Sects. 5.6 and 5.7), such as planetary-Rossby waves that can propagate upward. A remarkable manifestation of the effects of waves on the mean flow is encountered in the winter stratosphere, where the mean flow undergoes dramatic thermal and dynamical changes because of planetary wave breaking. Sudden stratospheric warmings (SSWs) are spectacular transient events in the winter Northern Hemisphere (NH) first discovered by Scherhag (1952). The winter polar temperature increases significantly within a few days following the breakdown/weakening of the stratospheric polar vortex as a consequence of rapid planetary wave amplification and breaking. The winter polar warming is accompanied by deceleration, and even reversals of the eastward (westerly) zonal mean zonal winds, $\bar{u}$, at 60°N at around 30 km (10 hPa). If $\bar{u}$ reverses its direction during a warming, then it is called a "major warming". If the mean flow is significantly weakened but not reversed during an appreciable warming at 90°N, then the warming is categorized as "minor" (Holton 1980). During a major warming, the temperature at the North Pole may increase up to 60 K.

The origin and characteristics of these events have been the subject of many studies. Matsuno (1971) has developed the first dynamical model of idealized SSW events and has demonstrated that planetary wave interactions with the stratospheric mean flow is responsible for the occurrence of SSWs. Further numerical studies confirmed Matsuno (1971)'s conclusion qualitatively (Holton 1976; Palmer 1981). Schoeberl (1978) has presented one of the earliest overviews of the theory and observation of stratospheric warmings focusing on the middle atmosphere.

Observed variations of atmospheric parameters during the 2008/2009 SSW are seen in Fig. 6.3 as presented in the work by Goncharenko et al. (2010). At the North Pole, the zonal mean temperature $\bar{T}$ increases from about 200 to 260 K within a few days. Averaged temperatures for the high-latitude sector between 60°N and 90°N demonstrate significant increase as well, while the zonal mean zonal wind is

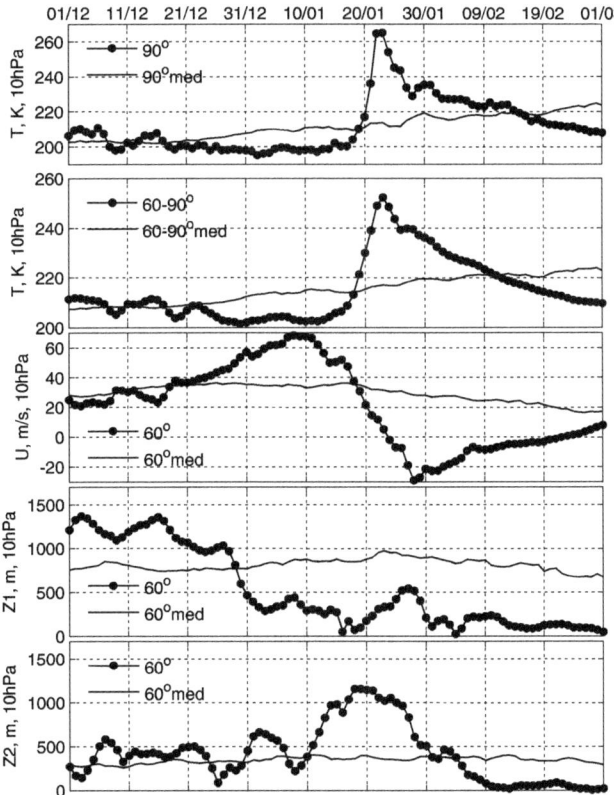

Fig. 6.3 Atmospheric parameters during a sudden stratospheric warming. Adopted from the work by Goncharenko et al. (2010, Fig. 1)

decelerated and reverses its direction. The associated planetary wave-1 (PW1) and planetary wave-2 (PW2) amplitudes presented in the lower two panels exhibit significant fluctuations. In this specific warming, PW2 amplitudes are enhanced while the PW1 is weakening before and during the warming. These observations indicate that PW2 plays a more important role for 2008–2009 major warming. Overall, an SSW is an example of transient mean flow forcing due to large-scale eddies. Each warming event can demonstrate different dynamical characteristics and the structure and evolution of stratospheric dynamics during warmings are quiet chaotic and impredictable. This variability is largely associated with nonlinear nature of wave transience/dissipation. In this context, wave transience means that the amplitude of a wave under consideration is changing with time. So, how does the mean flow return to its initial stage before the warming? Towards the final stages of the warming, it is seen that the mean flow is relatively easterly, which then inhibits wave vertical propagation to the stratosphere. Therefore, the wave-induced residual circulation $(\bar{v}^*, \bar{w}^*)$ decreases, radiative cooling processes gradually establishes the cold winter pole. Via the thermal wind relation, these conditions reestablish the westerly winter polar vortex.

The dynamical and thermal effects of warming are not confined to the stratosphere. Downward coupling to the troposphere can occur during warmings (Limpasuvan et al. 2004). Significant amount of vertical coupling between the lower and upper atmosphere can occur during warmings as well. The next section will specifically look at (gravity) wave-induced effects in the upper atmosphere during warmings.

6.10   Atmospheric Vertical Coupling and Gravity Waves During Sudden Warmings

Observations during SSW events demonstrate substantial changes in the circulation and thermal structure of the middle atmosphere (e.g., Manney et al. 2005; Thurairajah et al. 2010; Yuan et al. 2012; Zülicke and Becker 2013). However, the effects of these sudden warmings are not only local. A large number of publications demonstrate that significant amount dynamical and thermal effects can be detected at much higher altitudes in the mesosphere, thermosphere, and ionosphere (e.g., Goncharenko and Zhang 2008; Chau et al. 2009; Kurihara et al. 2010; Goncharenko et al. 2010, 2012; Pedatella and Forbes 2010; Pancheva and Mukhtarov 2011; Jonah et al. 2014). Incoherent scatter radar measurements have demonstrated an unprecedented view of the wave-like variations in the middle-latitude ionosphere during the 2010 sudden warming (Goncharenko et al. 2013). General circulation models are great tools to study SSW-induced changes in the atmosphere. Especially, because lower and the upper atmospheres can self-consistently be coupled within the framework of a GCM, the direct effects of the warming at higher levels can be quantified. For example, local time- and height-dependent response of the upper atmosphere has been simulated by the Japanese Ground-to-topside Model of Atmosphere and Ionosphere for Aeronomy (GAIA) (Liu et al. 2013).

If we want to couple the processes taking place in the lower atmosphere to upper layers, we first need to discuss the role of the stratosphere in the general circulation of the atmosphere. This region is of great importance in terms of vertical coupling. The primary source of lower atmospheric internal waves are situated in the bottom layers of the atmosphere, that is, below the stratosphere. The morphology of the stratosphere can greatly impact the upward propagation characteristics of these waves (Yiğit and Medvedev 2015). For example, gravity wave propagation and dissipation are highly variable because both the dynamical and thermal structure of the background atmosphere influence their upward propagation. For a given gravity wave, upward propagation to higher altitudes depends highly on the direction and magnitude of the wind. Specifically, the intrinsic phase speed of the ith harmonic, $c_i - \bar{u}$, determines to what extent the wave can propagate upward. During SSWs, for a fixed source of GW generation at lower levels, the stratospheric mean wind $\bar{u}$ changes rapidly and even changes the direction during major warming, such as the one shown for 2008/2009 winter in Fig. 6.3. Therefore, we expect substantial changes in the degree of GW penetration into the upper atmosphere because of the SSW-induced changes in the GW intrinsic parameters.

GCM simulations of how GW propagation into the thermosphere during a typical minor warming changes is shown in Fig. 6.4 as implemented from the work by Yiğit and Medvedev (2012, Fig. 2a). The extended gravity wave parameterization of Yiğit et al. (2008) is used to calculate gravity wave propagation and dissipation in the GCM and the GW activity is represented in terms of GW-induced root-mean-square (RMS) wind fluctuations. The vertical dashed line on 17 Feb denotes the onset of the minor warming after which GW activity in the thermosphere increases dramatically by a factor of three, indicating that GWs preferentially propagate into the thermosphere

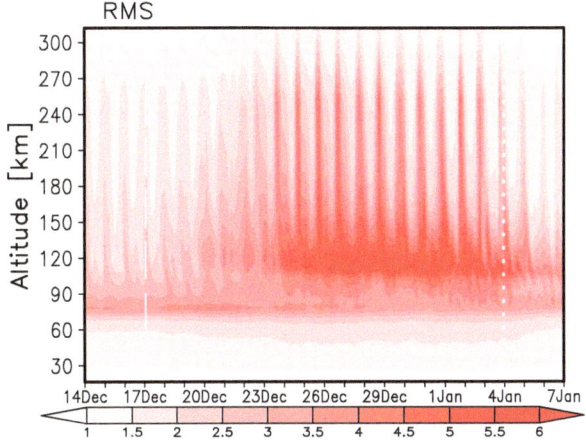

Fig. 6.4 Altitude-Universal time variations of gravity wave activity in terms of root-mean-square (RMS) wave fluctuations in m s^{-1} during a minor warming simulated with a general circulation model incorporating the extended gravity wave parameterization of Yiğit et al. (2008). Adopted from Yiğit and Medvedev (2012, Fig. 2)

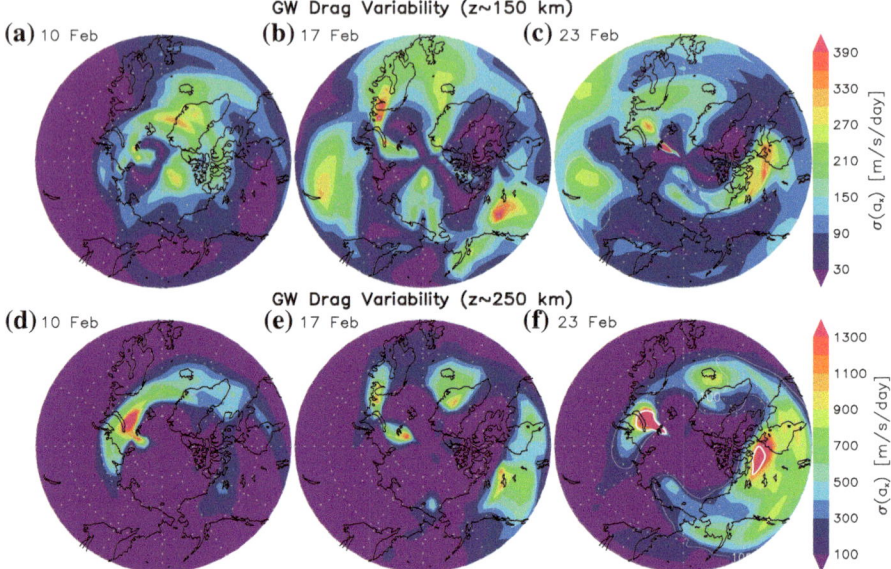

Fig. 6.5 Temporal variability of gravity wave effects during a minor sudden stratospheric warming simulated by a general circulation model implementing the extended parameterization of Yiğit et al. (2008). Adopted from (Yiğit et al. 2014, Fig. 3)

during the warming. Large GW activity in the thermosphere is sustained throughout the final stage of the warming on 4 Jan.

During SSWs GW mean effects undergo substantial enhancement. However, there is a significant degree of variability in the stratosphere during the warming. Therefore, one can predict an appreciable modulation of SSW-induced GW propagation and effects in the thermosphere. Yiğit et al. (2014) have investigated how the variability of GW zonal drag vary during the different phases of a minor warming. Figure 6.5 presents the temporal variability of GW drag at the initial stage, middle-stage and the peak phases of the warming at two representative thermospheric altitudes (150 and 250 km) during a simulated minor warming. There is a significant degree of temporal variability of GW drag in the lower and middle thermosphere of the warming and it varies substantially during all phases of the warming reaching a peak in the strongest phase of the warming. These results show in a spectacular way how the transient processes in the stratosphere can influence internal wave propagation to higher altitudes and can produce variability in the thermosphere at a range of altitudes at all phases of the warming. I expect that a possible interpretation of observed changes in the thermosphere-ionosphere during warming in the context of small-scale gravity wave effects will shed more light into SSW-induced vertical coupling in Earth's atmosphere.

References

Andrews DG, McIntyre ME (1976) Planetary waves in horizontal and vertical shear: the generalized eliassen-palm relation and the mean zonal acceleration. J Atmos Sci 33:2031–2048

Andrews DG, Holton JR, Leovy CB (1987) Middle atmosphere dynamics. International geophysics series, vol 40. Academic Press, London

Beagley SR, de Grandepre J, Koshyk JN, McFarlane NA, Shepherd TG (1997) Radiative-dynamical climatology of the first generation Canadian middle atmosphere model. Atmos Ocean 35:293–331

Becker E (2011) Dynamical control of the middle atmosphere. Space Sci Rev 168:283–314. doi:10. 1007/s11214-011-9841-5

Bjerknes V (1904) Das Problem von der Wettervorhersage, betrachtet vom Standpunkt der Mechanik und derPhysik. Meteor Z 21:1–7

Charney JG, Drazin PG (1961) Propagation of planetary-scale disturbances from the lower into the upper atmosphere. J Geophys Res 66(1):83–109

Chau JL, Fejer BG, Goncharenko LP (2009) Quiet variability of equatorial e×b drifts during a sudden stratospheric warming event. Geophys Res Lett 36:L05101. doi:10.1029/2008GL036785

Courant R, Friedrichs K, Lewy H (1928) über die partiellen Differenzengleichungen der mathematischen Physik. Mathematische Annalen 100(1):32–74. doi:10.1007/BF01448839, http://dx.doi. org/10.1007/BF01448839

Dunkerton T, Hsu CPF, McIntyre ME (1981) Some eulerian and lagrangina diagnostics for a model stratospheric warming. J Atmos Sci 38:819–843

Edmon HJ, Hoskins BJ, McIntyre ME (1980) Eliassen-palm cross sections for the troposphere. J Atmos Sci 37:2,600–2,616

Fritts DC (1984) Gravity wave saturation in the middle atmosphere: a review of theory and observations. Rev Geophys Space Phys 22:275–308

Fritts DC, Alexander MJ (2003) Gravity wave dynamics and effects in the middle atmosphere. Rev Geophys 41(1):1003. doi:10.1029/2001RG000106

Fritts DC, Vadas SL, Wan K, Werne JA (2006) Mean and variable forcing of the middle atmosphere by gravity waves. J Atmos Sol-Terr Phys 68:247–265

Fuller-Rowell TJ, Rees D, Quegan S, Moffett RJ, Codrescu MV, Millward GH (1996) A coupled thermosphere-ionosphere model (CTIM). In: Schunk RW (ed) Programme solar terrestrial energy (STEP) Handbook. Utah, pp 217–238

Garcia RR, Marsh DR, Kinnison DE, Boville BA, Sassi F (2007) Simulations of secular trends in the middle atmosphere. J Geophys Res 112:D09301. doi:10.1029/2006JD007485

Gardner LC, Schunk RW (2011) Large-scale gravity wave characteristics simulated with a high-resolution global thermosphere-ionosphere model. J Geophys Res 116:A06303. doi:10.1029/ 2010JA015629

Goncharenko L, Zhang SR (2008) Ionospheric signatures of sudden stratospheric warming: ion temperature at middle latitude. Geophys Res Lett 35:L21103. doi:10.1029/2008GL035684

Goncharenko LP, Coster AJ, Chau JL, Valladares CE (2010) Impact of sudden stratospheric warmings on equatorial ionization anomaly. J Geophys Res 115:A00G07. doi:10.1029/2010JA015400

Goncharenko LP, Coster AJ, Plumb RA, Domeisen DIV (2012) The potential role of stratospheric ozone in the stratosphere-ionosphere coupling during stratospheric warmings. J Geophys Res 39:L08101. doi:10.1029/2012GL051261

Goncharenko LP, Hsu VW, Brum CGM, Zhang SR, Fentzke JT (2013) Wave signatures in the midlatitude ionosphere during a sudden stratospheric warming of January 2010. J Geophys Res Spacc Phys 118: doi:10.1029/2012JA018251

Helmholtz H (1858) über Integrale der hydrodynamischen Gleichungen, welche den Wirbelbewegunen entsprechen. J fur die reine und angewandte Mathematik 25–55

Holton JR (1976) A semi-spectral numerical model for wave-mean flow interactions in the stratosphere: application to sudden stratospheric warmings. J Atmos Sci 33:1639–1649

Holton JR (1980) The dynamics of stratospheric warmings. Ann Rev Earth Planet Sci 8:169–190

Holton JR, Hakim GJ (2012) An introduction to dynamic meteorology, 5th edn. Academic Press, London

Holton JR, Wehrbein WM (1980) A numerical model of the zonal mean circulation of the middle atmosphere. Pure Appl Geophys 118:284–306

Jin H, Miyoshi Y, Fujiwara H, Shinagawa H, Terada K, Terada N, Ishii M, Otsuka Y, Saito A (2011) Vertical connection from the tropospheric activities to the ionospheric longitudinal structure simulated by a new Earth's whole atmosphere-ionosphere coupled model. J Geophys Res Space Phys 116(A1): doi:10.1029/2010JA015925, http://dx.doi.org/10.1029/2010JA015925

Jonah OF, de Paula ER, Kherani EA, Dutra SLG, Paes RR (2014) Atmospheric and ionospheric response to sudden stratospheric warming of January 2013. J Geophys Res Space Phys 119(6):4973–4980. doi:10.1002/2013JA019491, http://dx.doi.org/10.1002/2013JA019491

Kleppner D, Kolenkow RJ (1973) An introduction to mechanics. McGraw-Hill, New York

Kurihara J, Ogawa Y, Oyama S, Nozawa S, Tsutsumi M, Hall CM, Tomikawa Y, Fujii R (2010) Links between a stratospheric sudden warming and thermal structures and dynamics in the high-latitude mesosphere, lower thermosphere, and ionosphere. Geophys Res Lett 37:L13806. doi:10.1029/2010GL043643

Lamb H (1932) Hydrodynamics, 6th edn. Dover, originally published as Treatise on the mathermatical theory of the motion of fluids in 1879

Laštovička J (2006) Forcing of the ionosphere by waves from below. J Atmos Sol-Terr Phys 68:479–497

Liebermann R (1999) Eliassen-palm fluxes of the 2-day wave. J Atmos Sci 56:2846–2861

Limpasuvan V, Thompson DWJ, Hartmann DL (2004) The life cycle of the northern hemisphere sudden stratospheric warmings. J Clim 17:2584–2596

Liu H, Jin H, Miyoshi Y, Fujiwara H, Shinagawa H (2013) Upper atmosphere response to stratosphere sudden warming: local time and height dependence simulated by GAIA model. Geophys Res Lett 40: doi:10.1002/grl.50146

Manney GL, Krüger K, Sabutis JL, Sena SA, Pawson S (2005) The remarkable 2003–2004 winter and other recent warm winters in the Arctic stratosphere since the late 1990s. J Geophys Res 110:D04107. doi:10.1029/2004JD005367

Matsuno T (1971) A dynamical model of the stratospheric sudden warming. J Atmos Sci 28:1479–1494

Mengel JG, Mayr HG, Chan KL, Hines CO, Reddy CA, Arnold NF (1995) Equatorial oscillations in the middle atmosphere generated by small scale gravity waves. Geophys Res Lett 22:3027–3030

Miyoshi Y, Fujiwara H, Jin H, Shinagawa H (2014) A global view of gravity waves in the thermosphere simulated by a general circulation model. J Geophys Res Space Phys 119:5807–5820. doi:10.1002/A019848

Palmer TN (1981) Diagnostic study of a wavenumber-2 stratospheric sudden warming in a transformed euleraian-mean formalism. J Atmos Sci 38:844–855

Pancheva D, Mukhtarov P (2011) Stratospheric warmings: the atmosphere-ionosphere coupling paradigm. J Atmos Sol-Terr Phys 73:1697–1702

Pedatella NM, Forbes JM (2010) Evidence for stratosphere sudden warming-ionosphere coupling due to vertically propagating tides. Geophys Res Lett 37:L11104. doi:10.1029/2010GL043560

Richardson LF (1922) Weather prediction by numerical process. Cambridge University Press, Cambridge

Ridley AJ, Deng Y, Tóth G (2006) The global ionosphere-thermosphere model. J Atmos Sol-Terr Phys 68:839–864

Scherhag R (1952) Die explosionsartige Stratosphärenerwärmung des Spätwinters 1951–1952. Ber Deut Wetterdienstes 6:51–63

Schoeberl MR (1978) Stratospheric warmings: observation and theory. Rev Geophys 16(4):521–538

Schoeberl MR, Strobel DF (1978) The zonally averaged circulation of the middle atmosphere. J Atmos Sci 35:577–591

Smith AK (2012) Global dynamics of the MLT. Surv Geophys 33: doi:10.1007/s10712-012-9196-9

Smith RW (2000) The global-scale effect of small-scale thermospheric disturbances. J Atmos Sol-Terr Phys 62:1623–1628

Thurairajah B, Collins RL, Harvey VL, Lieberman RS, Gerding M, Mizutani K, Livingston JM (2010) Gravity wave activity in the Arctic stratosphere and mesosphere during the 2007–2008 and 2008–2009 stratospheric sudden warming events. J Geophys Res 115:D00N06. doi:10.1029/2010JD014125

Yiğit E, Medvedev AS (2012) Gravity waves in the thermosphere during a sudden stratospheric warming. Geophys Res Lett 39:L21101. doi:10.1029/2012GL053812

Yiğit E, Medvedev AS (2015) Internal wave coupling processes in Earth's atmosphere. Adv Space Res 55:983–1003. doi:10.1016/j.asr.2014.11.020, http://www.sciencedirect.com/science/article/pii/S0273117714007236

Yiğit E, Ridley AJ (2011) Role of variability in determining the vertical wind speeds and structure. J Geophys Res 116:A12305. doi:10.1029/2011JA016714

Yiğit E, Aylward AD, Medvedev AS (2008) Parameterization of the effects of vertically propagating gravity waves for thermosphere general circulation models: sensitivity study. J Geophys Res 113:D19106. doi:10.1029/2008JD010135

Yiğit E, Medvedev AS, England SL, Immel TJ (2014) Simulated variability of the high-latitude thermosphere induced by small-scale gravity waves during a sudden stratospheric warming. J Geophys Res Space Phys 119: doi:10.1002/2013JA019283

Yuan T, Thurairajah B, She CY, Chandran A, Collins RL, Krueger DA (2012) Wind and temperature response of midlatitude mesopause region to the 2009 sudden stratospheric warming. J Geophys Res 117:D09114. doi:10.1029/2011JD017142

Zülicke C, Becker E (2013) The structure of the mesosphere during sudden stratospheric warmings in a global circulation model. J Geophys Res Atmos 118: doi:10.1002/jgrd.50219

Appendix A
Physical Constants and Parameters

A.1 Physical Constants

In Table A.1 the physical constants that are used in this book are summarized.

A.2 Notation of Physical Parameters

Table A.2 summarized a list of major physical parameters used in the book.

Table A.1 List of physical constants used in the book in alphabetical order

Constant	Label	Value	Dimension
Astronomical unit	AU	149.6×10^6	km
Atomic mass unit	u	1.66×10^{-27}	kg
Avogadro's number	N_A	6.02×10^{23}	$mole^{-1}$
Boltzmann constant	k_b	$1.38 \times^{-23}$	$J\ K^{-1}$
Gravitational constant	G	6.67×10^{-11}	$N\ m^2\ kg^{-2}$
Speed of light in vacuum	c	2.99×10^8	$m\ s^{-1}$
Stefan-Boltzmann constant	σ_b	5.67×10^{-8}	$W\ m^{-2}\ K^{-4}$
Universal gas constant	R	8.317	$J\ mole^{-1}\ K^{-1}$

Approximate values of the constants are stated

© The Author(s) 2015
E. Yiğit, *Atmospheric and Space Sciences: Neutral Atmospheres*,
SpringerBriefs in Earth Sciences, DOI 10.1007/978-3-319-21581-5

Table A.2 Below is a list of important physical parameters used in the book.

Notation	Parameter	Dimension
c	Phase speed	m s^{-1}
$\hat{c}$	Intrinsic phase speed	m s^{-1}
c_p	Specific heat copacity at constant pressure	J K^{-1} kg^{-1}
c_v	Specific heat copacity at constant volume	J K^{-1} kg^{-1}
D	Diffusion coefficient	m^2 s^{-1}
D	Diffusion tensor	
f	Coriolis parameter $(= 2\Omega \sin\theta)$	rad s^{-1}
f_0	Midlatitude Coriolis parameter	rad s^{-1}
F	Eliassen-Palm flux vector	
H	Scale height	m
k	Wave vector	
K	Wave number	m^{-1}
L	Angular momentum	kg m^2 s^{-1}
n	Number of moles	
N	Buoyancy frequency	s^{-1}
q	Heat flux	J s^{-1} m^{-2}
Q	Heat	J
p	Pressure	kg m^{-1} s^{-2}
p	Momentum	kg m s^{-1}
r	Position vector	m
$\Re$	Reynolds number	
R_i	Richardson number	
t	Time	s
T	Temperature	K
u	Three-dimensional neutral velocity	m s^{-1}
u	Zonal speed	m s^{-1}
v	Meridional speed	m s^{-1}
w	Vertical speed	m s^{-1}
$\bar{v}^*$	Meridional residual mean velocity	m s^{-1}
$\bar{w}^*$	Vertical residual mean velocity	m s^{-1}
U	Internal energy	J
v$_i$	Three-dimensional ion velocity	m s^{-1}
V	Volume	m^3
ε	Internal energy per unit mass	m^3
ρ	Mass density	kg m^{-3}

(continued)

Table A.2 (continued)

Notation	Parameter	Dimension
θ_T	Potential temperature	K
ν	Specific volume ($= \rho^{-1}$)	$m^3\ kg^{-1}$
	Kinematic viscosity ($\mu\rho^{-1}$)	$m^2\ s^{-1}$
Φ	Particle flux	$m^{-2}\ s^{-1}$
	Geopotential	$m^2\ s^{-2}$
ω	Angular frequency	$rad\ s^{-1}$
ω	Vorticity	s^{-1}
Ω	Planetary rotational angular frequency	$rad\ s^{-1}$
Ω	Planetary rotation vector	$rad\ s^{-1}$

This is not a full list of parameters

Appendix B
Useful Mathematical Tools

B.1 Integration

The integration of a function $f(x)$ in a definite integral $[A, B]$ is given by

$$\int_A^B f(x)dx = F(x = B) - F(x = A), \qquad (B.1)$$

where $\frac{dF(x)}{dx} = f(x)$ and A and B are the lower and upper limits, respectively.

B.2 Spherical Coordinate System

The Spherical coordinate map S maps a point P in (r, ϕ, θ)-space to a point in (x, y, z)-space:

$$S : \begin{pmatrix} r \\ \phi \\ \theta \end{pmatrix} \mapsto \begin{pmatrix} x = r \cos \phi \cos \theta \\ y = r \sin \phi \cos \theta \\ z = r \sin \theta \end{pmatrix}, \qquad (B.2)$$

where r is the radial distance, ϕ is the longitude, and θ is the latitude. Longitude and latitude are measured in radians, where 2π radians is equal to $360°$. The "del" or "nabla" operator is given by

$$\nabla \equiv \frac{\partial}{\partial r}\hat{\mathbf{r}} + \frac{1}{r}\frac{\partial}{\partial \theta}\hat{\boldsymbol{\theta}} + \frac{1}{r\cos\theta}\frac{\partial}{\partial \phi}\hat{\boldsymbol{\phi}}, \qquad (B.3)$$

© The Author(s) 2015
E. Yiğit, *Atmospheric and Space Sciences: Neutral Atmospheres*,
SpringerBriefs in Earth Sciences, DOI 10.1007/978-3-319-21581-5

where $r \ll 0$, $0 \leq \theta \leq \pi$, and $0 \leq \phi \leq 2\pi$. The position vector can therefore be written as

$$\mathbf{r}(r, \phi, \theta) = r \cos \phi \cos \theta \, \hat{\mathbf{i}} + r \sin \phi \cos \theta \, \hat{\mathbf{j}} + r \sin \theta \, \hat{\mathbf{k}} \tag{B.4}$$

and the vector line element is

$$d\mathbf{r}(r, \phi, \theta) = \frac{\partial \mathbf{r}}{\partial r} dr + \frac{\partial \mathbf{r}}{\partial \phi} d\phi + \frac{\partial \mathbf{r}}{\partial \theta} d\theta \tag{B.5}$$

Glossary

Atmosphere-Ionosphere system Neutral and plasma environment of Earth that extends from the ground to the top of the ionosphere at an altitude of 500–1000 km.

Conservation laws Fundamental laws of nature stating that certain quantities, such as, energy, momentum, charge, and mass, have the unique property that they can be transported within the medium (or universe in general). Therefore, the total amount of a conserved quantity is constant, however, it can be redistributed within a given conserving system.

Eddy Small-scale turbulent vortex structures in geophysical fluids. It is also used to represent small rotational features in physical parameters, e.g., eddy currents.

Eulerian mean Zonal averaging, that is, averaging around a longitude circle.

General circulation The large-scale circulation of the atmosphere described by the totality of the motion of neutral air.

General circulation models First principle three-dimensional mathematical models of the coupled differential equations of atmospheric dynamics and thermodynamics that solve the time-dependent evolution of atmospheric energy, mass, and momentum distributions.

Geostrophic balance An approximation to the equation of motion of atmospheric dynamics, in which the pressure gradient force balances the Coriolis force. The associated flow structures are the geostrophic winds.

Gravity waves These are atmospheric waves that are primarily generated with a broad spectrum in the lower atmosphere and are able to propagate upward. These are an obiquitous feature of all planetary atmospheres.

Navier-Stokes equations The set of nonlinear dynamical equations governing the motion of an air parcel in planetary atmospheres.

Radiation Transport of energy by electromagnetic wave propagation.

Space weather A collective expression for the effects of the Sun on Earth atmosphere and geospace environment.

Sudden stratospheric warming A stratospheric event during which the neutral temperature at the North Pole increases by tens of Kelvin, accompanied by the weakening of the stratospheric zonal mean eastward jets around 60°N.

© The Author(s) 2015   107
E. Yiğit, *Atmospheric and Space Sciences: Neutral Atmospheres*,
SpringerBriefs in Earth Sciences, DOI 10.1007/978-3-319-21581-5

Sunspot   Cold and dark area of large magnetic field strength on the surface of the sun.

Vertical coupling   Exchange of energy and momentum, and mass transport between the different atmospheric layers in the vertical direction.

Vortex   Microscopic description of rotation in fluids.

Index

A
Alfvén waves, 45
Angular velocity, 32
Atmosphere-ionosphere, 46

B
Boussinesq approximation, 61
Buoyancy frequency, 60

C
Conduction, 25
    heat flux, 25
Correlation function, 90
Courant-Friedrichs-Lewy, 87

D
Diffusion, 23
Dispersion relation, 57, 62

E
Eddy, 82, 88
Ensemble average, 90
Eulerian mean, 89
Exosphere, 49
Extended gravity wave parameterization, 64

F
Finite difference technique, 86

G
GCM, 71, 86

General circulation, 85
    models, 86
Geostrophic balance, 37
Gravitational constant, 14
Gravity waves, 60
Group velocity, 58

H
Heat capacity, 18
    specific heat ratio, 18
Heliosphere, 43

I
Ideal gas law, 16
Incompressibility, 62
Incompressible flow, 35
Intrinsic phase speed, 58
Ionosphere, 50

L
Laplace operator, 24
Log-pressure system, 38

M
Magnetosphere, 45
    magnetopause, 45
Mean flow, 88
Mesosphere, 48

N
Navier-Stokes equations, 36

© The Author(s) 2015
E. Yiğit, *Atmospheric and Space Sciences: Neutral Atmospheres*,
SpringerBriefs in Earth Sciences, DOI 10.1007/978-3-319-21581-5